W9-CFK-978

Modern MOS Technology:

Processes, Devices, and Design

Modern MOS Technology:
Processes, Devices, and Design

DeWITT G. ONG

Program Manager, Intel Corporation
Chandler, Arizona

McGraw-Hill Book Company

New York   St. Louis   San Francisco
Auckland   Bogotá   Hamburg
Johannesburg   London   Madrid   Mexico
Montreal   New Delhi   Panama   Paris
São Paulo   Singapore   Sydney   Tokyo
Toronto

Library of Congress Cataloging in Publication Data

Ong, DeWitt G.
  Modern MOS technology.

  Includes index.
  1. Metal oxide semiconductors. 2. Integrated circuits. I. Title. II. Title: Modern
M.O.S. technology.
TK7871.99.M44053   1984   621.381′73042   83-11339
ISBN 0-07-047709-4

Copyright © 1984 by McGraw-Hill, Inc. All rights reserved. Printed in the United States
of America. Except as permitted under the Copyright Act of 1976, no part of this
publication may be reproduced or distributed in any form or by any means, or stored in a
data base or retrieval system, without the prior written permission of the publisher.

1234567890   KGP/KGP   89876543

ISBN 0-07-047709-4

The editors for this book were Harry Helms and Joan Cipriano, the designer was Elliot
Epstein, and the production supervisor was Thomas G. Kowalczyk. It was set in
Baskerville by University Graphics, Inc.

Printed and bound by The Kingsport Press.

To My Parents,
Dr. and Mrs. Tiong Giok Ong,
and My Wife,
Josephine

Contents

Preface   **xi**

1   OVERVIEW OF MOS   **1**

1-1   Introduction, 1
1-2   What Is a MOS Transistor?, 1
1-3   Characteristics of a MOS Transistor, 5
1-4   MOS Transistor Transfer Characteristics, 6
1-5   Comparison between MOS and Bipolar Transistors, 7

2   REVIEW OF SEMICONDUCTOR PHYSICS   **2**

2-1   Crystalline Nature of Silicon, 11
2-2   Nature of Intrinsic Silicon, 13
2-3   Doping with Donors, 16
2-4   Doping with Acceptors, 18
2-5   Electrical Conductivity, 19
2-6   *pn* Junction at Zero Bias, 20
2-7   *pn* Junction with Forward Bias, 22
2-8   *pn* Junction with Reverse Bias, 24
2-9   The Diode Equation, 24
2-10 Electric Field and Breakdown Voltage of *pn* Junctions, 26
2-11 Depletion-Layer Capacitance, 28

3   SURFACE PROPERTIES OF SILICON   **33**

3-1   Energy-Band Diagram for the Ideal Case, 33
3-2   Calculation of the Threshold Voltage V_T, 36
3-3   Nonideal Effects, 39

4   *CV* PLOTS   **49**

4-1   The Importance of *CV* Plots, 49
4-2   High-Frequency *CV* Plots, 49
4-3   Low-Frequency *CV* Plots, 51
4-4   Equations for *CV* Plots, 53
4-5   Normalization of the Ideal *CV* Plot, 55
4-6   Deep Depletion, 57
4-7   Deviations from the Ideal *CV* Plots, 58
4-8   Practical Considerations in Doing *CV* Plots, 67
4-9   *CV* Analysis Program, 70
4-10   Measurements of Minority Carrier Lifetime, 71
Appendix   *CV* Analysis Program, 78

5   MOS DEVICE PHYSICS   **81**

5-1   Triode Region, 81
5-2   Saturation Region, 84
5-3   Avalanche Region, 86
5-4   Subthreshold Region, 88
5-5   Second-Order Effects, 90
5-6   Ways of Measuring Threshold Voltage, 95
5-7   MOS Device Applications, 99

6   SMALL-GEOMETRY EFFECTS   **109**

6-1   Introduction, 109
6-2   Nonuniform Doping and Effect on Threshold Voltage, 109
6-3   Subthreshold Current, 111
6-4   Short-Channel Effect, 114
6-5   Narrow-Width Effect, 122
6-6   Small-Geometry Effects, 124
6-7   Shrink and Scaling, 125
6-8   Scaling Down the Dimensions of MOS Devices, 127
6-9   Summary, 129

7   MOS IC PROCESSES   **133**

7-1   Metal-Gate PMOS, 133
7-2   The Hypothetical Metal-Gate NMOS, 135
7-3   Metal-Gate CMOS, 136
7-4   Silicon Gate LOCOS NMOS Process, 138
7-5   The HMOS Process, 142
7-6   Process Enhancements, 144

8   MOS WAFER FABRICATION   **149**

8-1   Oxidation, 149
8-2   Diffusion, 155

8-3   Ion Implantation, 162
8-4   Deposition, 172
8-5   Photolithography, 175
8-6   Etch, 179
8-7   Clean, 182

9   MOS DIGITAL IC DESIGN   **187**

9-1   Building Blocks for MOS Digital IC, 187
9-2   Inverter DC Analysis, 189
9-3   Inverter Transient Analysis, 195
9-4   MOS Logic Circuits, 197
9-5   Memory Circuits, 207
9-6   Other Circuit Techniques, 216

10   ANALOG MOS DESIGN   **223**

10-1   Considerations in Analog MOS Circuits, 223
10-2   Analog Building Blocks, 227
10-3   MOS Operational Amplifiers, 233
10-4   Capacitor-Based Circuits, 237
10-5   Switched-Capacitor Filters, 239
10-6   Charge-Coupled Devices, 245
10-7   Charge-Coupled Device Applications, 251

11   CMOS   **261**

11-1   Advantages and Disadvantages of CMOS, 261
11-2   CMOS Circuits, 262
11-3   CMOS Circuit Analysis, 265
11-4   Power Dissipation, 266
11-5   Processing Issues, 268
11-6   Latch-Up, 270
11-7   Silicon on Sapphire, 276

12   COMPUTER CIRCUIT SIMULATION   **279**

12-1   Need for Circuit Simulation Programs, 279
12-2   ASPEC Circuit Simulation Program, 280
12-3   Transient Analysis Sample Run, 281
12-4   AC Analysis Sample Run, 283
Appendix   ASPEC User's Manual, Version 7.0 (Abridged), 300

13   LAYOUT, MASK, AND ASSEMBLY   **317**

13-1   Layout and Design Rules, 317
13-2   From Logic Design to Masks, 321
13-3   Automated Chip Layout, 327
13-4   Assembly, 331

14 YIELD AND RELIABILITY   339

14-1 Importance of Yield, 339
14-2 Yield Models, 340
14-3 Reliability and Failure Models, 344
14-4 Burn-In, 351

Table of Physical Constants   **357**

Table of Material Properties   **357**

Table of Conversion Constants   **358**

Solutions to Selected Problems   **358**

Index   **361**

Preface

Metal-oxide-semiconductor (MOS) integrated circuit (IC) technology encompasses a wide spectrum of sciences and disciplines, ranging from semiconductor physics to chemistry to circuit design. The goal of this book is to distill the more important concepts and techniques of the various disciplines and present them in a coherent framework. As stated by the subtitle, this body of information is partitioned into device physics (Chaps. 1 to 6), wafer processing (Chaps. 7 and 8), and circuit design (Chaps. 9 to 14).

Chapter 1 gives an overview of the MOS technology and its importance in the IC industry. Chapter 2 reviews semiconductor physics with an emphasis on the pn junction for its similarity in using the concept of band bending. Threshold voltage is introduced in Chap. 3 as part of the discussion on the surface properties of silicon. A separate chapter (Chap. 4) is devoted to CV (capacitance-voltage) plots, not only because of their importance as an analysis tool in MOS work but also because they are a vehicle for discussing nonideal effects. Included in the chapter are practical tips in doing CV plots and a computer program (written in BASIC) for analyzing the plots. Chapter 5 covers physics and characteristics of MOS transistors, and Chap. 6 probes the subject further with regard to small geometry effects.

Chapter 7 describes the various MOS processes that are in volume production now or will be in the foreseeable future. Chapter 8 describes the individual processing steps in wafer fabrication. Attention is focused on the newer processing techniques such as ion implantation, dry etching, and noncontact exposure.

The section on circuit design starts off in Chap. 9 with digital IC design, the mainstay of the MOS business. Chapter 10 covers analog MOS design, an area of growing importance and interest. Switched capacitors and charge-coupled devices (CCDs) are the two main methods of implementation.

CMOS is discussed separately in Chap. 11 because it is increasingly specified as the technology to use, particularly for larger scale integration. Chapter 12 uses the commercially available ASPEC circuit simulation program to provide one specific example of computer design aids. Chapter 13 explains the rest of the design and manufacturing cycle, namely layout, mask generation, and assembly. Finally, the book closes in Chap. 14 with yield and reliability, two key subjects inseparable from any study of integrated circuits.

For the convenience of the reader, each chapter is designed to be self-contained, with minimum referencing back and forth to other sections of the book. However, the order of the chapters is found from experience to be pedagogically effective for easy understanding and long-term retention. The book strives for logical progression from the familiar to the unfamiliar, from the understanding of a device, to its fabrication, then to its use.

Each chapter is written at a level appropriate for the intended reader. For example, the review of semiconductor physics is presented in very basic terminology for those with no background in that area, whereas the section on digital circuit design assumes familiarity with logic manipulations.

This book uses charts, graphs, and tables to convey specific technical information useful to practicing professionals in the field. At the same time, it is suitable as a textbook for a one-semester course. It is appropriate at either the senior or first-year graduate level, depending on the depth of coverage desired. As an aid in teaching, there are problem sets at the ends of the chapters. However, they are really an integral part of the book. They amplify the materials in the text, illustrate numerical values, reinforce the explanations, and introduce additional concepts.

I am indebted to Professor Lex Akers of Arizona State University for writing Chap. 6, "Small-Geometry Effects," a subject of timely importance.

I am also indebted to many reviewers for their enlightening comments and discussions on various materials in the book. They include Dr. J. R. Brews and Dr. J. A. Cooper, Jr., of Bell Laboratories; Dr. E. T. Gaw, Dr. W. W. Fisher, Dr. S. T. Nieh, Dr. M. H. Woods, Mr. F. J. Burris, and Dr. J. Mar of Intel Corporation; Professors P. R. Gray and R. W. Brodersen of University of California in Berkeley; and Mr. S. L. Smith, Mr. M. Smith, Mr. R. Allgood, and Dr. J. B. Price of Motorola Incorporated. I am very appreciative of the work of Mrs. Kathy Gear in typing the whole manuscript on the word processor and of Mr. Eric Maass in generating a solution set for the problems.

Last, I would like to thank my wife, Josephine, and my children. Their encouragement and patience during the many evenings and weekends that went into the writing of this book make them the unseen coworkers of this writing project.

DeWitt G. Ong

1

Overview of MOS

1-1 INTRODUCTION

Metal-oxide semiconductor (MOS) integrated circuits (ICs) have become the dominant technology in the semiconductor industry, overtaking the bipolar integrated circuits in sales volume in recent years (see Fig. 1-1).[1] With MOS, it is now possible to have more than 100,000 transistors on a single chip, allowing the fabrication of a complete 32-bit microprocessor[2] with a three-chip set (see Fig. 1-2) or a quarter of a million bits of memory on a chip.[3] The main reason behind MOS IC's pervasiveness is that MOS ICs exceed the bipolar transistors in functional density, that is, the number of functions performed on a single chip, and at a competitive price as well. That, in turn, is due to the fact that MOS transistors are smaller in size and somewhat simpler to fabricate.

At this point, it should be noted that by and large, MOS products are sold almost exclusively in integrated-circuit form. Very few single transistors are marketed, and those are primarily specialized power transistors and RF mixer transistors. By contrast, in the bipolar semiconductor market, there is a sizable demand for discrete transistors and diodes.

1-2 WHAT IS A MOS TRANSISTOR?

As the name metal-oxide semiconductor implies, the MOS transistor consists of semiconductor material (silicon) on which is grown a thin layer (250 to 1000 Å) of insulating oxide, topped by a gate electrode. Such a structure is shown in Fig. 1-3. The gate electrode was originally metal, specifically aluminum, but is now more commonly a layer of polycrystalline silicon (referred to as *polysilicon*). Source and drain junctions are formed with a

1

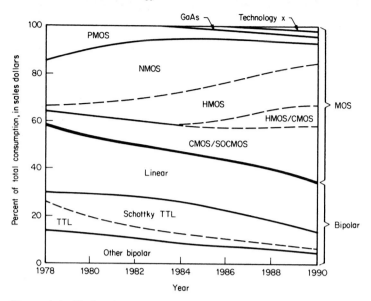

Figure 1-1 **Market share for different semiconductor technologies.** *(After Ref. 1. Copyright 1983. ICE Corporation.)*

small overlap with the gate. A generic term for such a device might be IGFET, which stands for insulated gate field-effect transistor. Though the two terms are used interchangeably, MOSFET is used more commonly.

In the discussion to follow, and throughout the rest of the book, we shall use n-channel transistors for our analysis. The analysis for the p-channel transistor follows simply by invoking duality; i.e., all semiconductor types are reversed, as well as the polarity of applied voltages.

In the n-channel transistor of Fig. 1-3, the semiconductor body (or substrate) is p-type and the source and drain diffusions are n-type. With no voltage applied to the gate, there is no conduction path between the source and drain because they are merely back-to-back diodes. When a voltage, positive with respect to the substrate, is applied to the gate and is of such a magnitude that it is greater than a certain threshold voltage V_T, then electrons are attracted to the surface of the semiconductor. (Note that electrons are minority carriers in the p-type substrate.) In fact, so many electrons are attracted to the surface that an extremely thin (approximately 50-Å) channel is formed, where the semiconductor actually changes from p- to n-type. Now, when a voltage is applied between source and drain, current can flow. The current flow is via majority carriers, since electrons flow through all n-type materials. (Bipolar transistor current carriers are minority carriers in the base region; they are affected by recombination phenomenon.) Transistor gain results from the ability of the gate voltage to modulate the channel conductivity. The electrons pile up at the oxide-semiconductor interface (i.e.,

Figure 1-2 The instruction decoder unit of Intel 432, a 32-bit microprocessor. It contains 110,000 transistors on a die size of 320 mils on a side. (*Intel Corp.*)

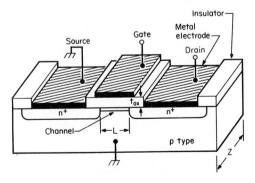

Figure 1-3 An *n*-channel MOS transistor. (*After S. M. Sze, Physics of Semiconductor Devices. 2d ed., John Wiley and Sons, Chap. 8, 1981.*)

3

the silicon surface) rather than flow through the gate circuit because the gate oxide prevents any dc gate current from flowing.

The device just described requires a voltage to be applied before the channel is formed and is called an *enhancement-mode transistor* (see Fig. 1-4). Sometimes, when there are built-in positive charges at the oxide-semiconductor interface or when charges are deliberately introduced by ion implantation, a channel will be formed even when no gate voltage is applied. Such a transistor is called a *depletion-mode transistor,* because a reverse polarity (in this case, negative) gate voltage would have to be applied to deplete the

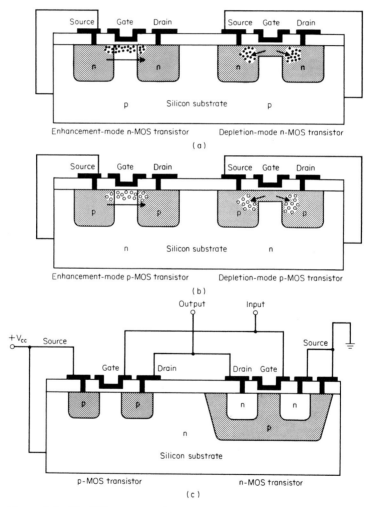

Figure 1-4 The different types of MOS transistors: (a) NMOS, (b) PMOS, and (c) CMOS. *(From J. D. Meindl, Scientific American, pp. 70–81, Sept. 1977.)*

channel of electrons and shut the device off. Its threshold voltage is thus negative. Note that the depletion-mode *n*-channel devices still conduct via electrons.

Transistors such as that of Fig. 1-3 which conduct via electrons are called *n-channel devices* (NMOS). One can build a similar set of *p*-channel (PMOS) devices that conduct via holes by reversing all voltage polarities and semiconductor types. That is, source and drain are *p*-type on *n*-type substrate, and the threshold voltage is negative for enhancement devices and positive for depletion devices, as shown in Fig. 1-4.

Notice that MOS transistors are always four-terminal devices. In the discussion so far, and in most applications, the substrate is often tied to the source, but one must always be conscious of tying the substrate (fourth terminal) some place.

If one were to fabricate PMOS and NMOS devices on the same substrate to build complementary MOS (CMOS) circuits, there would be a basic inconsistency in substrates. This is solved by using a *p*-tub diffusion to create the background for the *n*-channel devices, as shown in Fig. 1-4. Again note where the fourth terminal for each device is tied.

There are many symbols in use for MOS transistors. The more common ones are listed in Fig. 1-5 with set (*a*) being the most widely used. A good aid for remembering the symbols is that the arrow always points toward the *n* region, be it the *n* channel or the *n*-type semiconductor.

1-3 CHARACTERISTICS OF A MOS TRANSISTOR

Some of the characteristics of a MOS transistor are

1. *Bilaterally Symmetric*   This means that the source and drain are electrically interchangeable. In NMOS, the more positive of the two is the drain. This is contrasted with bipolar transistors where the transistor gain would be severely degraded if the emitter and collector leads were interchanged.

2. *Unipolar*   MOS transistors conduct exclusively via one type of carrier. This contrasts with bipolar transistors which, although they conduct primarily via one type of carrier (e.g., electrons in *npn*), do have both types of carriers flowing at the same time.

3. *High-Input Impedance*   Because of the gate oxide, there is no dc path between the gate and the other terminals. The input impedance is then extremely high, and is primarily capacitive. Its dc resistance is greater than 10^{14} Ω.

4. *Voltage-Controlled*   MOS devices are voltage-controlled. When that is coupled with the fact of high-input impedance, the result is a device

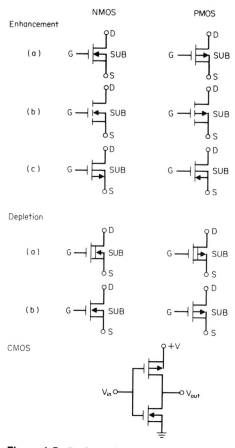

Figure 1-5 Logic symbols for MOS.

with extremely low input power. One MOS transistor can then drive many other transistors similar to it, i.e., it has high fan-out capability. Bipolar transistors, on the other hand, are current-controlled devices.

5. *Self-Isolating* MOS ICs can be very dense because the MOS transistors are self-isolating. The drain of one device is naturally isolated from the drain or source of the others by means of back-to-back diodes. This eliminates the need for the deep, and hence also wide, isolation diffusions in bipolar processes.

1-4 MOS TRANSISTOR TRANSFER CHARACTERISTICS

If one ties the substrate to the source, then the transfer characteristics (I_D versus V_D) of a MOS transistor would look like Fig. 1-6. For a given applied gate voltage that is greater than the threshold voltage V_T, the drain

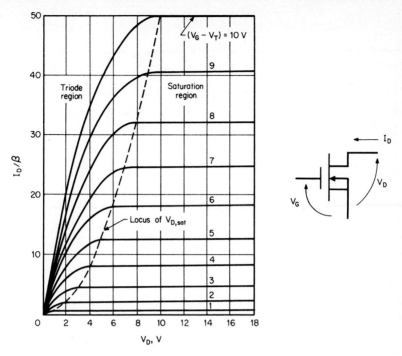

Figure 1-6 Idealized transfer characteristics (I_D versus V_D) of a MOS transistor.

current will rise linearly with increasing drain-to-source voltage. However, the rate of rise decreases until the drain current soon saturates to a constant value. At a higher gate voltage, the same shaped curve will result except the current values will be higher.

There are two regions in the transfer characteristics plot. The first, where current is rising, is the triode (or linear) region, where the drain current is described by

(1-1) $I_D = \beta[(V_G - V_T)V_D - \tfrac{1}{2}V_D^2]$

where β is a constant. The second region is the saturation region where the drain current does not depend on V_D:

(1-2) $I_D = \tfrac{1}{2}\beta(V_G - V_T)^2$

The point where the two lines meet is at the saturation drain voltage:

(1-3) $V_D = V_{D,\text{sat}} = V_G - V_T$

1-5 COMPARISON BETWEEN MOS AND BIPOLAR TRANSISTORS

MOS transistors are generally smaller than bipolar transistors. Figure 1-7 shows a comparison between single MOS and bipolar transistors in terms

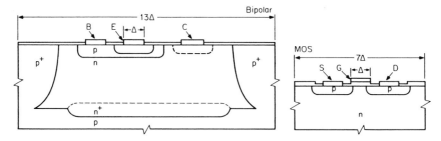

Figure 1-7 **Layout comparison of a bipolar transistor and a MOS transistor.** *(After Ref. 4.)*

of size. For any given processing resolution Δ, the bipolar transistor will require a larger area because of the additional alignment tolerances from having three separate emitter, base, and collector masking and etching layers. But what really uses up the largest area is the isolation diffusion. In order to penetrate the deep n-epitaxial layer, the p^+ isolation diffusion has to spread laterally by approximately 70 percent of the layer depth. Recently, the use of an oxide trench to replace the isolation diffusion has increased the density of bipolar circuits significantly. And by using integrated injection logic (I^2L) for digital circuits, the density is increased to that approaching MOS. This is accomplished by operating the transistor in the inverse gain mode and using the n-epitaxial layer as emitter, so it can be made common to all transistors in the circuit.

REFERENCES

1. Integrated Circuit Engineering Corporation, *Status '83—A Report on the Integrated Circuit Industry,* Scottsdale, Arizona, 1983.

2. W. S. Richardson, J. A. Bayliss, et al., "The 32b computer instruction decoding unit," *Int. Solid State Circuit Conf. Proc.,* pp. 114–115 and 262, Feb. 18, 1981.

3. S. Nakajima, K. Kiuchi, et al., "1 μm 256K RAM process technology using molybdenum-polysilicon gate," *Int. Electr. Dev. Meeting Proc.,* Dec. 7, 1981.

4. R. M. Warner, "Comparing MOS and bipolar integrated circuits," *IEEE Spectrum,* p. 50, June 1967.

PROBLEMS

1. What do we mean when we say MOSFETs are
   (a) Bilaterally symmetric?
   (b) Unipolar?

2. An n-channel enhancement-mode MOSFET conducts via (holes or electrons) and has a (positive or negative) threshold voltage. When the threshold voltage is made

(negative or positive) to form a depletion-mode device, e.g., by ion implantation, the device will then conduct via (electrons or holes).

3. Plot the MOSFET device equation

$$I_D = \beta[(V_G - V_T)V_D - \tfrac{1}{2}V_D^2]$$

for

$$V_G - V_T = 10 \text{ V}$$

$$\beta = 1 \text{ A/V}^2$$

and V_D in steps of 2 V from 0 to 20 V. Compare your plot with the output characteristics in Fig. 1-6.

4. If the gate of a transistor is tied to its drain, what region of operation would it be in (saturation or triode)? Draw its two-terminal VI characteristics.

2

Review of
Semiconductor Physics

2-1 CRYSTALLINE NATURE OF SILICON

Silicon as utilized in integrated circuits is crystalline in nature. As with all crystalline material, silicon consists of a repeating basic unit structure called a *unit cell*. For silicon, the unit cell consists of an atom surrounded by four equidistant nearest neighbors which lie at the corners of a tetrahedron, enclosed by dashed lines in Fig. 2-1. However, the whole structure of Fig. 2-1 can be considered as two meshed face-centered cubic unit cells with one displaced by one-quarter of the length down the body diagonal (Fig. 2-2). Its significance lies in the fact that since a cube has three axes of symmetry, silicon, in bulk form, is isotropic. That is, all of the physical properties, e.g., mobility, resistivity, etc., are the same when measured in any direction on bulk material. Those same properties would not necessarily still be isotropic if they were measured on thin films of silicon, or if *surface* properties are of interest. Surface properties would depend on the orientation of the crystal since, for one thing, the number of atoms exposed per unit surface area would be different.

In MOS work, it is often necessary to specify various crystallographic planes or surfaces in a silicon crystal. This is done by specifying the *Miller indexes* as follows:

1. Determine the three intercepts a, b, c of the plane with the three axes X, Y, Z, respectively.

2. The reciprocals of these intercepts comprise the set of three numbers enclosed in parentheses that form the Miller indexes:

$$\left(\frac{1}{a,} \frac{1}{b,} \frac{1}{c,} \right)$$

11

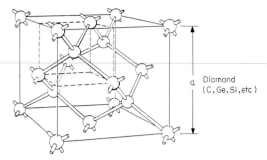

Figure 2-1   Tetrahedron structure of silicon crystal.

Figure 2-2   Face-centered cubic unit cell.

The three numbers must be reduced to the smallest set of integers. As examples, the (111) and (100) planes are shown in Figs. 2-3 and 2-4. In the case of the (100) plane the intercepts are at 1, ∞, ∞. The reciprocal of these values yields (100) as the Miller index.

MOS integrated circuits are fabricated from wafers that are sawed parallel to either the (100) or (111) plane, with (100) materials being the most common by far. This is due largely to the fact that when (100) wafers are oxidized during MOS processing, they yield the lowest residual charge at the oxide-semiconductor interface (see Chap. 3).

By convention, when parentheses enclose the Miller indexes, they refer to the given plane. Square brackets define the direction, and angle brackets

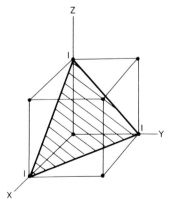

Figure 2-3   The (111) plane.

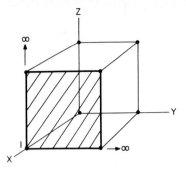

Figure 2-4 The (100) plane.

refer to the equivalent set of directions, that is, $\langle 1, 0, 0 \rangle = [1, 0, 0]$, $[0, 1, 0]$, $[0, 0, 1]$.

2-2 NATURE OF INTRINSIC SILICON

Silicon that is free of doping impurities is called *intrinsic*. Silicon has a valence of 4 and forms covalent bonds with four other neighboring silicon atoms. This sharing of electrons between neighboring atoms is represented by the two-dimensional cross section in Fig. 2-5.

A very important concept in the study of semiconductors is the energy-band diagram (see Fig. 2-6). It is used to represent the range of energy a valence electron can have. For semiconductors such as silicon, it is known that electrons can have any one value of a continuous range of energy levels while they occupy the valence shell of the atom.[1] That band of energy levels is called the *valence band*. Within the same valence shell, but at a slightly higher energy level, is yet another band of continuously variable, allowed energy levels. This is the conduction band. Between these two bands is a range of energy levels where there are no allowed states for an electron. This

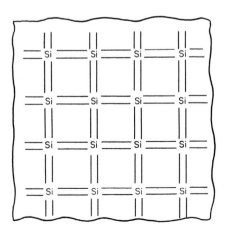

Figure 2-5 Covalent bonding of silicon atoms.

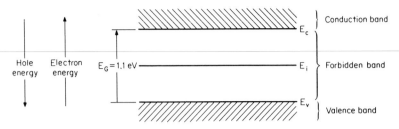

Figure 2-6   **Energy-band diagram.**

is the band gap E_G. In silicon, E_G = 1.1 eV (electron volt) at room temperature. The energy-band diagram of Fig. 2-6 simply plots electron energy increasing in the $+y$ direction. In Chap. 3, it is shown that energy levels can bend as a function of distance into the silicon. In that case, the x direction has significance. By convention the highest level of the valence band is marked E_v, the lowest level of the conduction band is called E_c, and the level midpoint in between is defined as E_i.

Notice that there is no zero electron energy level indicated in the energy-band diagram. This reflects the fact that the absolute energy level is not as useful as changes in energy levels. All equations involving the energy-band diagram will always show differences in energy levels, and not a single energy level.

At absolute zero temperature, the valence band is completely filled with valence electrons and no current flow can take place. But at any temperature above absolute zero, a small number of electrons possess enough thermal energy to be excited into the conduction band and can react to applied electric fields to conduct current. These are the mobile electrons. On the other hand, for every electron excited into the conduction band, a vacancy is left in the valence band, creating a hole. When an electric field is applied, a nearby electron can move in to fill that empty state, creating another hole. The hole then appears to move in response to the applied field with a certain hole mobility.

This intrinsic concentration of holes and electrons is relatively small for silicon at room temperature, resulting in silicon being a semiconductor. But it doubles roughly every 8 to 10°C. The temperature dependence is given by[2]

$$(2\text{-}1) \quad n = p = n_i = 3.9 \times 10^{16}\, T^{3/2} \exp\left(\frac{-E_G}{2kT}\right) \quad \text{cm}^{-3}$$

where n = electron concentration, cm^{-3}
p = hole concentration, cm^{-3}
n_i = intrinsic carrier concentration, cm^{-3}
T = temperature, K

E_G = band gap, eV (1.21 eV for silicon at 0 K)

k = Boltzmann's constant

At T = 300 K,

(2-2) $n_i = 1.5 \times 10^{10}$ cm^{-3}

Insulating materials have a large E_G, for example, ~9 eV for SiO_2, and extremely few electrons would gain enough thermal energy to jump the band gap. Metals, on the other hand, have overlapping bands and therefore do not have band gaps. The valence electrons that are near the top of the filled region of the band can easily gain thermal energy and become mobile electrons. When an external field is applied, metals then act highly conductive.

The Fermi level is most commonly used in semiconductor work to describe the energy level of the electronic state at which one electron has a probability of 0.5 of occupying. But the Fermi level also describes the number of electrons and holes. With reference to the energy-band diagram in Fig. 2-7, we find that

(2-3) $n = n_i \exp \dfrac{E_F - E_i}{kT}$

and

(2-4) $p = n_i \exp \dfrac{E_i - E_F}{kT}$

For intrinsic material,

(2-5) $p = n = n_i$

Thus

(2-6) $E_F = E_i$

In other words, for the intrinsic material, the Fermi level is at the midpoint of the band gap. This explains the origin of the symbol E_i to designate the midpoint.

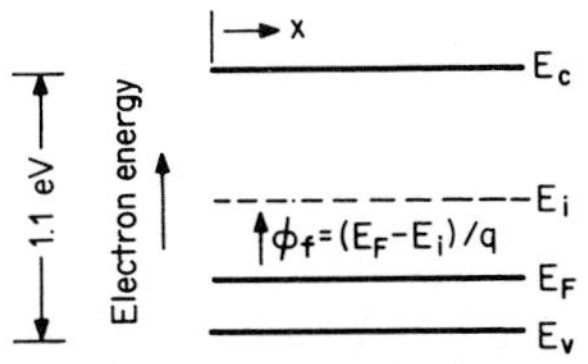

Figure 2-7 Energy-band diagram with Fermi level shown.

More detailed analysis shows that E_i, the Fermi level for intrinsic silicon, is actually displaced from midgap by a term that is very small (a few kT).

2-3   DOPING WITH DONORS

Semiconductors that are doped with impurities are classified as *extrinsic*. The first type of extrinsic silicon to be discussed is n-type silicon.

If an atom with a valence of 5, such as phosphorus, replaces a silicon atom in a substitutional manner, then only four electrons are used to share in the covalent bonding, and there would be an excess electron (Fig. 2-8). Even at a low temperature, thermal energy can easily overcome the ionization energy, and this electron breaks free and becomes available for current conduction. In terms of the energy-band diagram (Fig. 2-9), the extra electron of phosphorus lies in a "shallow" state just below the conduction band edge E_C and can easily be excited by thermal energy into the conduction band. The phosphorus atom is a *donor* of an electron and becomes a positively charged ion, but the silicon material as a whole still remains electrically neutral. Note that the donor atom is physically immobile and is definitely not a *hole*. This concept of mobile vs. immobile charge is quite important and will be invoked again in later discussions on MOS structures.

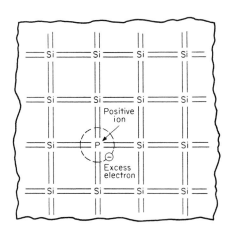

Figure 2-8   Covalent bonding of donor (phosphorus) doped silicon.

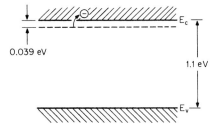

Figure 2-9   Phosphorus donor energy level in silicon.

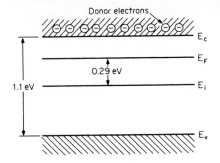

Figure 2-10 Energy-band diagram for silicon with donor (phosphorus) doping concentration of 10^{15} cm^{-3}. $T = 300$ K.

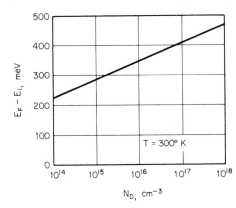

Figure 2-11 $E_F - E_i$ vs. donor doping concentration N_D.

At room temperature, practically all phosphorus atoms are fully ionized and have contributed an electron each. Since the contribution from thermal generation ($n = n_i$) is negligible when compared with typical doping concentration, then the electron concentration is

(2-7) $\quad n = N_D^+$

where N_D is the doping concentration per cubic centimeter. The plus sign is simply a reminder that the donor atom itself is positively charged. The hole concentration, though neglected in Eq. (2-7), is not zero but rather

(2-8) $\quad p = \dfrac{n_i^2}{n}$

It is suppressed far below the intrinsic value ($p = n_i$).

Knowing the electron concentration from Eq. (2-7) allows one to determine the Fermi level by means of the defining equation, Eq. (2-3). Figure 2-10 shows the Fermi level for n-type material with a typical doping concentration of $N_D = 10^{15}$ cm^{-3}. Figure 2-11 shows the Fermi level (with reference to E_i) for other donor doping concentrations.

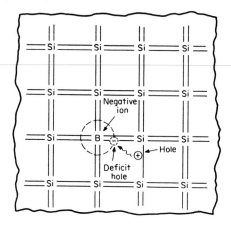

Figure 2-12 Covalent bonding of acceptor (boron) doped silicon.

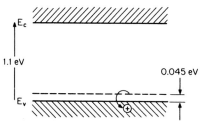

Figure 2-13 Boron acceptor energy level in silicon.

2-4 DOPING WITH ACCEPTORS

The doping of silicon with acceptors to form p-type silicon is analogous to the n-type silicon situation. An acceptor, such as boron, has a valence of 3 and readily accepts an electron from the valence band, leaving behind a hole for conducting current (see Fig. 2-12). In terms of the energy-band diagram in Fig. 2-13, the extra energy level of boron lies on a "shallow" level just above the valence band edge E_v. An electron is easily excited from the valence band, leaving behind a mobile hole. The boron atom itself is negatively charged (having accepted an electron) but is immobile and not, of itself, available for current conduction. To maintain charge neutrality in the p-type material, the hole concentration is

$$(2\text{-}9) \quad p = N_A^-$$

where N_A is the acceptor doping concentration in centimeters^{-3} and the minus sign is a reminder that the boron is negatively charged. The minority carrier is suppressed to a level of

$$(2\text{-}10) \quad n = \frac{n_i^2}{p}$$

The Fermi level for p-type material is below E_i and is illustrated in Fig. 2-14 for a typical acceptor doping concentration of $N_A = 10^{15}$ cm^{-3}. For other acceptor doping concentrations, one can use Fig. 2-11 by simply substituting N_A for N_D and reversing the sign of $E_F - E_i$.

2-5 ELECTRICAL CONDUCTIVITY

Conduction of electricity through bulk silicon is primarily through its majority carrier. For n-type material, the majority carriers are electrons. The conductivity of n-type material is

(2-11) $\quad \sigma_n = nq\mu_n$

where $\sigma_n = (\Omega \cdot \text{cm})^{-1}$
$\qquad n$ = electron concentration, cm^{-3}
$\qquad q$ = electronic charge
$\qquad \mu_n$ = electron mobility, cm^2/V$\cdot$s

The resistivity is the inverse,

(2-12) $\quad \rho_n = \dfrac{1}{nq\mu_n}$

For p-type material, the conductivity and resistivity are analogous:

(2-13) $\quad \sigma_p = pq\mu_p$

(2-14) $\quad \rho_p = \dfrac{1}{pq\mu_p}$

The mobility described above is the proportionality constant between the applied electric field and resultant carrier drift velocity

(2-15) $\quad v = \mu E$

The mobility referred to in Eqs. (2-11) to (2-15) is the bulk mobility. It is shown in subsequent chapters that mobility is an important parameter for

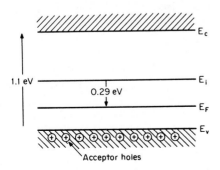

Figure 2-14 Energy-band diagram for silicon with an acceptor (boron) doping concentration of 10^{15} cm^{-3}.

MOS devices; however, those discussions refer to surface mobility, that is, mobility of a carrier constrained to travel along a semiconductor surface. Figure 2-15 shows that bulk electron mobility is approximately three times hole mobility. Both decrease with increasing doping concentration[3] because of increased collisions with charged impurities. Mobility also generally decreases with increasing temperature because of increased scattering by thermal vibration of the lattice.

By taking into account the fact that mobility changes with doping concentration, and utilizing Eqs. (2-12) and (2-14), one can reconstruct Irvin's curve[4] (Fig. 2-16) which plots resistivity as a function of doping concentration.

2-6 *pn* JUNCTION AT ZERO BIAS

An idealized *pn* junction is one where a uniformly doped *p*-type material abruptly changes to *n*-type as in Fig. 2-17*a*. A more practical structure, however, is one where one impurity is diffused into the other. Figure 2-17*b* shows one such example. When that junction is formed, electrons diffuse out from the *n*-type material and cross over the junction into the *p*-type region, leaving behind uncovered, positively ionized donor atoms. At the same time, holes diffuse out from the *p*-type silicon, exposing negatively ionized acceptor atoms. Enough of the mobile carriers will cross the junction

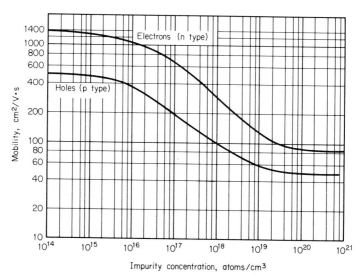

Figure 2-15 **Bulk carrier mobility as a function of doping concentration. Valid for both majority and minority carriers. ($T = 300$ K).**

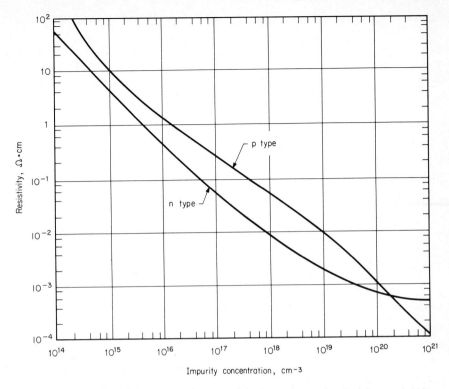

Figure 2-16 **Resistivity as a function of doping concentration (Irvin's curve).** *(After Ref. 4.)*

until the field set up by the ionized donors and acceptors on the two sides of the junction balances out the diffusion force, as in Fig. 2-18. (The center region that is depleted of mobile carriers is called the *depletion region*.) The potential difference set up by that field is the built-in potential. Take note that this built-in potential cannot do any external useful work, not even activating a voltmeter. (Why?)

The built-in potential is proportional to the amount of band bending in the energy-band diagram and is defined as

$$(2\text{-}16) \quad V_B \equiv \frac{E_{in} - E_{ip}}{q}$$

Equation (2-16) is another way of saying that V_B is the voltage of the p-type with respect to n-type side. V_B is also equal to the sum of the electron and hole Fermi potentials which we will define here for later use in Chap. 3 on MOS energy-band bending

$$(2\text{-}17) \quad V_B = -(\phi_{Fn} + |\phi_{Fp}|)$$

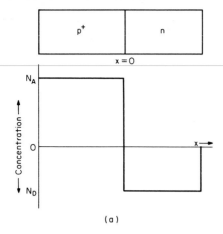

(a)

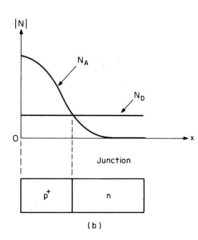

(b)

Figure 2-17 (*a*) Ideal *pn* junction; (*b*) practical *pn* junction.

where the electron *Fermi potential* is

(2-18) $\phi_{Fn} \equiv \dfrac{E_F - E_{in}}{q} = \dfrac{kT}{q} \ln \dfrac{N_D}{n_i}$

and the hole Fermi potential is

(2-19) $\phi_{Fp} \equiv \dfrac{E_F - E_{ip}}{q} = -\dfrac{kT}{q} \ln \dfrac{N_A}{n_i}$

2-7   *pn* JUNCTION WITH FORWARD BIAS

When a positive voltage V_A is applied to the *p* side of a junction, the *pn* junction is forward-biased. The Fermi level of the *p* side is displaced downward on the energy-band diagram by an amount qV_A (see Fig. 2-19). In effect, the built-in potential is now reduced, causing the diffusion current of majority carriers to flow across the depletion region. The carrier flow shows up as an increase in the minority carrier concentration on the other side of the junction (see Fig. 2-20). These excess minority carriers recombine rapidly outside of the depletion layer, falling to the equilibrium value within a diffusion length L.

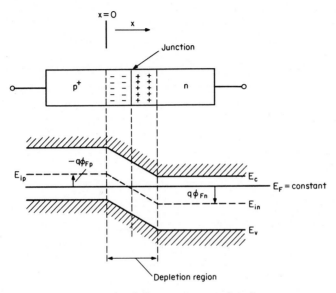

Figure 2-18 Energy-band diagram near a *pn* junction.

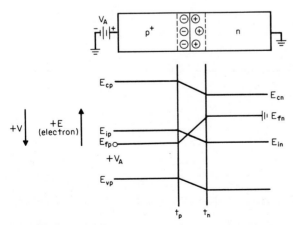

Figure 2-19 Energy-band diagram of a *pn* junction with forward bias.

Since the junction is no longer in equilibrium, then

(2-20) $pn \neq n_i^2$

The change in majority carrier concentration is so small that it does not show up in the logarithmic plot of Fig. 2-20.

2-8 *pn* JUNCTION WITH REVERSE BIAS

When the voltage applied to the *p*-type side is negative, the junction is reverse-biased. The applied voltage V_A adds directly to the built-in potential and causes the Fermi level of the *p* side E_{Fp} to be pushed up on the energy-band diagram by the amount qV_A (Fig. 2-21). The net result is that, this time, minority carriers are drawn toward the junction. There is now a decay of minority carriers, and this also takes place within a diffusion length L as in Fig. 2-22. The reverse-bias current is extremely small because the source of the current flow, the minority carrier concentration, is small.

2-9 THE DIODE EQUATION

The diode current of a *pn* junction can be shown to be

(2-21) $I = I_o \left[\exp\left(\dfrac{qV_A}{kT}\right) - 1 \right]$

where

(2-22) $I_o = qA \left(\dfrac{D_p p_{no}}{L_p} + \dfrac{D_n n_{po}}{L_n} \right)$

where A = diode area, cm^2
q = electronic charge

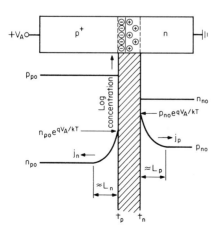

Figure 2-20 Carrier concentration for forward-biased *pn* junction.

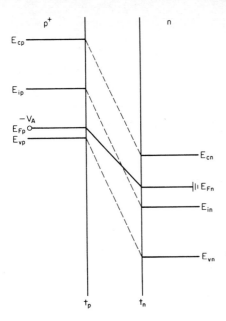

Figure 2-21 Energy-band diagram for *pn* junction with reverse bias.

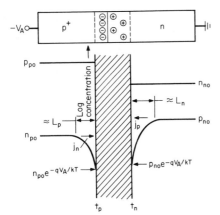

Figure 2-22 Carrier concentration for the reverse-biased *pn* junction.

D_p, D_n = hole and electron diffusion coefficients, cm^2/s
P_{no}, n_{po} = equilibrium minority carrier concentrations
L_p, L_n = hole and electron diffusion lengths

V_A is the applied voltage, defined with the *n*-type side of the junction as ground. With forward bias, V_A is positive and the exponential term dominates; hence

$$(2\text{-}23) \quad I_F = I_o \exp\left(\frac{qV_A}{kT}\right)$$

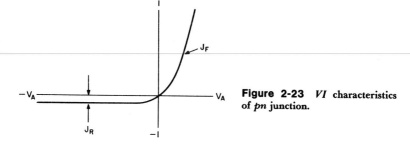

Figure 2-23 *VI* characteristics of *pn* junction.

With reverse bias, V_A is negative; then for most values of V_A,

(2-24) $\quad \exp\left(\dfrac{qV_A}{kT}\right) \ll 1$

Then

(2-25) $\quad I_R = -I_o$

The reverse current of an ideal diode is then constant with respect to reverse bias.

The voltage-current characteristics of the diode equation are plotted in Fig. 2-23. Equation (2-21) describes the reverse characteristics of germanium diodes quite well. But for silicon diodes, the leakage current is found to increase with increasing reverse bias because of the thermal generation current.

(2-26) $\quad I_{\text{gen}} = A\,\dfrac{n_i x_d}{2\tau}$

where x_d is the depletion width, and τ is the effective minority carrier lifetime, typically $\approx 10^{-6}$ s. For silicon diodes of practical areas, I_{gen} is typically $\leq 10^{-9}$ A; hence this dependence on reverse bias might not show up on most curve tracers.

Under strong illumination, a diode in reverse bias would generate a photocurrent that behaves in many ways similar to thermally generated current, except that it is not bias-dependent. Because of the strong possibility of photocurrent coming into play, it is good practice to always keep devices being tested in the dark.

2-10 ELECTRIC-FIELD AND BREAKDOWN VOLTAGES OF *pn* JUNCTIONS

Poisson's equation relates the electric field E to the charge density ρ in the depletion layer as

(2-27) $\quad \dfrac{dE(x)}{dx} = \dfrac{\rho(x)}{\epsilon_s \epsilon_o}$

where ϵ_s is the semiconductor dielectric constant, and ϵ_o is the permittivity of free space, 8.86×10^{-14} F/cm. In other words, the electric field is the integral of the charge distribution as a function of distance. For a step junction where one side is heavily doped, the electrical properties are determined by the characteristics of the lighter doped side. With the p^+n step junction example, the maximum field is, integrating from depletion edge x_d to the junction,

$$(2\text{-}28) \quad E_{\max} = \int_{-x_d}^{0} \frac{qN_D \, dx}{\epsilon_s \epsilon_o} = \frac{qN_D x_d}{\epsilon_s \epsilon_o}$$

To solve for x_d, we note that

$$(2\text{-}29) \quad E = -\frac{dV}{dx} = \frac{-qN_D x}{\epsilon_s \epsilon_o}$$

Integrating once, the potential difference across this one-sided step junction is found to be

$$(2\text{-}30) \quad V = \frac{-qN_D}{2\epsilon_s \epsilon_o} x_d^2 = V_B$$

V_B is numerically negative when the n-type side is the reference ground potential. Therefore

$$(2\text{-}31) \quad x_d = \sqrt{\frac{2\epsilon_s \epsilon_o}{qN_D} |V_B|}$$

With applied bias,

$$(2\text{-}32) \quad x_d = \sqrt{\frac{2\epsilon_s \epsilon_o}{qN_D} |V_B + V_A|}$$

As mentioned earlier, in reverse bias, V_B and V_A are of the same algebraic sign. Substituting Eq. (2-32) into (2-28) gives

$$(2\text{-}33) \quad E_{\max} = \sqrt{\frac{2qN_D}{\epsilon_s \epsilon_o} |V_B + V_A|}$$

When this maximum field exceeds a critical field of about 5×10^5 V/cm, then the diode breaks down and large current starts to flow. Rather than deal with critical field, a more useful quantity is the breakdown voltage. Figure 2-24 shows the breakdown voltage as a function of concentration.[5] However, Fig. 2-24 is for a plane junction. In silicon integrated circuits, junctions are formed by diffusion, with a certain curvature at the edges. That sharp curvature results in regions of enhanced field, thus lowering the breakdown voltage. The significance of this effect is evident in Fig. 2-25 where breakdown voltages are seen to be lowered when junctions have sharper curvature.

2-11 DEPLETION-LAYER CAPACITANCE

Under reverse-bias conditions a *pn* junction exhibits a voltage-dependent capacitance. To obtain this relationship, note that capacitance per unit area is

(2-34) $$C = \frac{dQ}{dV}$$

where dQ is the incremental change in charge due to an incremental change in voltage dV. For a step junction

(2-35) $$Q = qN_D x_d$$

Thus

(2-36) $$\frac{dQ}{dx_d} = qN_D$$

while

(2-37) $$V = \frac{qN_D x_d^2}{2\epsilon_s \epsilon_o}$$

Then

(2-38) $$\frac{dV}{dx_d} = \frac{qN_D x_d}{\epsilon_s \epsilon_o}$$

Therefore

(2-39) $$C = \frac{dQ/dx_d}{dV/dx_d} = \frac{qN_D}{qN_D x_d/\epsilon_s \epsilon_o} = \frac{\epsilon_s \epsilon_o}{x_d}$$

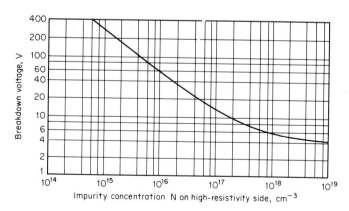

Figure 2-24 Breakdown voltage of a silicon step junction (n^+p or p^+n) as a function of doping concentration in the lightly doped side.

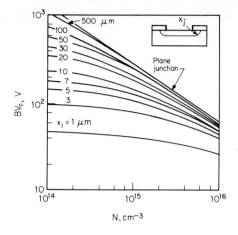

Figure 2-25 Breakdown voltage of planar, one-sided step junction with various radii of curvature.

This states that the junction capacitance is equivalent to that of a parallel plate capacitor separated by distance x_d in a dielectric medium of silicon. This agrees with our heuristic reasoning that with an incremental increase in charge caused by an increase in voltage, the depletion layer has to widen, uncovering an incremental amount of charge at the depletion layer's edge. Since, from Eq. (2-32),

$$(2\text{-}40) \quad x_d = \sqrt{\frac{2\epsilon_s\epsilon_o |V_A + V_B|}{qN_D}}$$

Therefore

$$(2\text{-}41) \quad C = \sqrt{\frac{\epsilon_s\epsilon_o qN_D}{2|V_A + V_B|}}$$

REFERENCES

1. R. B. Adler, A. C. Smith, and R. L. Longini, *Introduction to Semiconductor Physics,* Chaps. 1 and 2, John Wiley & Sons, New York, 1964.

2. W. N. Carr and J. P. Mize, *MOS/LSI Design and Application,* Chap. 1, McGraw-Hill, New York, 1972.

3. S. M. Sze and J. C. Irvin, "Resistivity, mobility, and impurity levels in GaAs, Ge, and Si at 300°K," *Solid-State Electron.,* vol. 11, p. 599, 1968.

4. J. C. Irvin, "Resistivity of bulk silicon and of diffused layers in silicon," *Bell Systems Tech. J.,* vol. 41, p. 387, 1962.

5. S. M. Sze, *Physics of Semiconductor Devices,* Chap. 3, Wiley, New York, 1969.

6. S. M. Sze and G. Gibbons, "Effects of junction curvature on breakdown voltages in semiconductors," *Solid-State Electron.*, vol. 9, p. 831, 1966.

PROBLEMS

1. Develop the expression for built-in voltage V_B that takes into account the doping concentration on both sides of the junction.

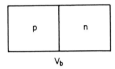

V_b

Figure 2-26 Problem 3.

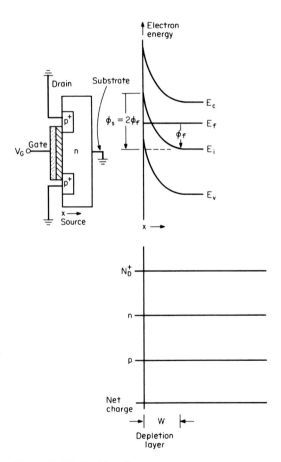

Figure 2-27 Problem 4.

2. True or false: A diode's reverse current does not depend on applied voltage as long as $|qV_A/kT| \gg 1$. Justify your answer.

3. Indicate polarity of built-in potential V_B shown in Fig. 2-26.

4. Sketch the carrier concentration profile for the energy-band diagram of a MOS transistor shown in Fig. 2-27. No numbers needed; simply sketch in relative magnitudes and rough profiles.

5. A silicon wafer is doped with 5×10^{15} boron atoms per cubic centimeter. At room temperature,
   (a) What is the hole concentration?
   (b) What is the electron concentration?
   (c) What is the resistivity of the wafer?

3

Surface Properties of Silicon

3-1 ENERGY-BAND DIAGRAM FOR THE IDEAL CASE

The concept of the energy-band diagram has been introduced in Chap. 2 with a note that it is drawn along a single vertical axis only. Then in the case of the *pn* junction, it was drawn as a function of distance *x*. In the case of a MOS structure, the energy band also changes as a function of distance, but the direction *x* is now defined to be from the gate into the silicon; therefore, in order to maintain the similarity with the way band diagrams were drawn before, the MOS transistor should be turned on its side, as in Fig. 3-1. The origin for *x* is the oxide-semiconductor interface. In an ideal case, and with no voltage applied, the energy bands would be perfectly flat. Hence this condition is called *flatband*. The concentrations of holes and electrons do not vary throughout the semiconductor. When voltage is applied to the gate, with substrate as ground, there will be bending of the bands in ways that are dependent on both the polarity of the voltage and the substrate type. (When source and drain are grounded, they are not involved in band bending.)

In the case of the *p*-type bulk as shown, a negative bias will attract mobile holes to the surface, making the material effectively more heavily doped *p* type. This is the *accumulation* condition as shown in Fig. 3-2*a*. When bias is changed to positive, the mobile holes are repelled from the surface, forming a *depletion* layer near the surface as in Fig. 3-2*b*. As the positive bias is increased, a point is reached when the oxide-semiconductor interface becomes *intrinsic* ($E_F = E_i$). Beyond that point, a special condition is reached when the surface potential, the amount of band bending, reaches

$$(3\text{-}1) \quad \phi_s = \frac{-(E_i \text{ at surface} - E_i \text{ at bulk})}{q} = 2\phi_f$$

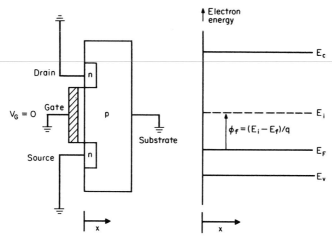

Figure 3-1 Energy-band diagram for gate region of a MOS transistor, ideal case.

where ϕ_f is the Fermi potential defined in Chap. 2. This is the *inversion* condition (Fig. 3-2c) and is of particular interest because a layer of charge is now formed that can be used for conduction in a MOS transistor. The gate voltage that brings about this condition is called the *threshold voltage*. This conduction layer is formed when a large number of mobile electrons have been attracted to the surface, either via thermal electron-hole pair generation or via the source junction from an external circuit. In addition, the number of electrons is large enough such that the carrier concentration at the surface is the same as the bulk, but of the opposite type. If the gate voltage is increased beyond the threshold voltage (which is actually the normal mode of operation), the band bending does not increase significantly (see Fig. 3-3). This comes from the fact that any slight further increase in the band bending past $2\phi_f$ increases the number of inversion-layer carriers exponentially, and thus the increased gate charge due to the additional bias is easily balanced.

In the case of an *n*-type bulk, the same sequence of events takes place, but with the polarity of voltages reversed. With positive gate voltage, majority carriers (electrons) *accumulate* at the surface making the silicon more *n* type. With negative gate voltage, a *depletion* layer is exposed as electrons are pushed back from the surface. Then at *inversion,* the depletion layer reaches a maximum width, and any further increases in gate bias are balanced by the inversion-layer charge comprised of mobile holes.

Figure 3-4 shows the charge distribution under the various biasing conditions for a *p*-type substrate MOS structure. Figure 3-4a shows the negative gate charge balanced by a hole sheet charge in accumulation. Figure 3-4b shows that in depletion, the gate charge is balanced by the depletion

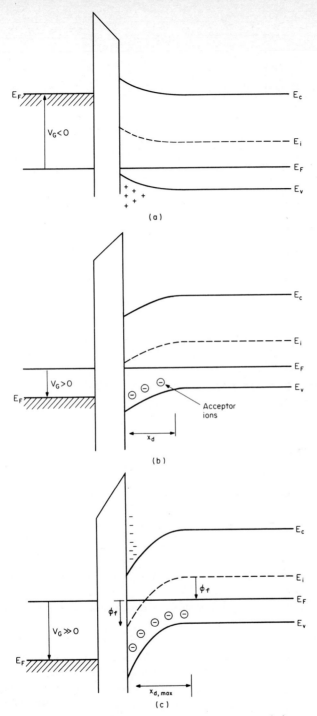

Figure 3-2 Band bending in *p*-bulk silicon: (*a*) accumulation; (*b*) depletion; (*c*) inversion.

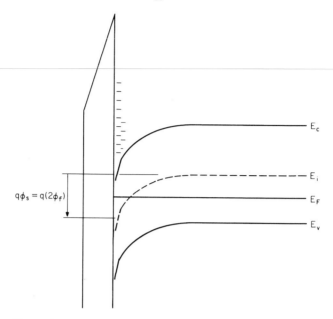

$q\phi_s = q(2\phi_f)$

E_c

E_i

E_F

E_v

Figure 3-3 **Band bending in *p*-bulk silicon in heavy inversion.**

layer; while Fig. 3-4c shows that in strong inversion, the gate charge is balanced by the inversion charge and depletion-layer charge.

Further discussion of the field effect of semiconductors can be found in Refs. 1 and 2.

3-2 CALCULATION OF THRESHOLD VOLTAGE V_T

Inversion is achieved when the surface potential, i.e., surface band bending, reaches

(3-2) $\phi_s = 2\phi_f$

However, it is more practical to work with the threshold voltage V_T, the *gate voltage* that is necessary to bring about this condition. An expression for V_T will now be derived.

An applied gate voltage V_G becomes the sum of voltage drops across the oxide and across the silicon:

(3-3) $V_G = V_{ox} + \phi_s$

where V_{ox} is the voltage across the gate oxide, and ϕ_s is the surface potential, equivalent to the voltage drop across the depletion layer. The goal is to express V_{ox} in terms of ϕ_s, which can then simply be set to $2\phi_f$. By definition

(3-4) $V_{ox} = E_{ox}t_{ox}$

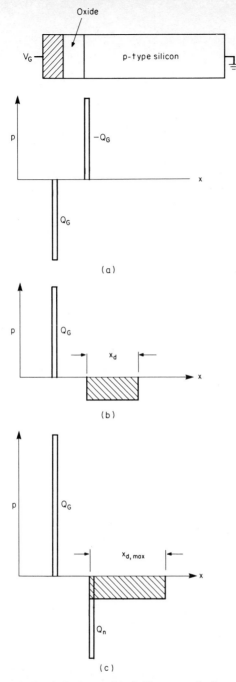

Figure 3-4 Energy-band diagram and charge distribution under various bias conditions for a *p*-type substrate (*n*-channel) MOS structure: (*a*) accumulation; (*b*) depletion; (*c*) inversion.

where E_{ox} is the electric field across the oxide of thickness t_{ox}. However, the continuity of the displacement vector D across the oxide-semiconductor interface requires that

(3-5) $\quad E_{ox}\epsilon_{ox} = E_{Si}\epsilon_s$

Furthermore, Gauss' theorem states that the field into the silicon surface is related to the amount of charge in the silicon by

(3-6) $\quad E_{Si} = \dfrac{-Q_s}{\epsilon_s\epsilon_o}$

where Q_s is the charge density (coulombs per square centimeter) in the silicon. In depletion Q_s is simply the depletion-layer charge. But in inversion it also includes the inversion-layer charge.

From Eqs. (3-4), (3-5), and (3-6) is obtained

(3-7) $\quad V_{ox} = \dfrac{-Q_s t_{ox}}{\epsilon_{ox}\epsilon_o} = -\dfrac{Q_s}{C_o}$

where the gate capacitance per unit area is defined as

(3-8) $\quad C_o = \dfrac{\epsilon_{ox}\epsilon_o}{t_{ox}}$

Therefore Eq. (3-3) is now

(3-9) $\quad V_G = \dfrac{-Q_s}{C_o} + \phi_s$

In depletion Q_s equals the depletion-layer charge Q_B

(3-10) $\quad Q_s = Q_B = -qN_A X_d = -\sqrt{2\epsilon_s\epsilon_o qN_A\phi_s}$

The last equality in Eq. (3-10) comes from the fact that the depletion-layer width, similar to Eq. (2-32), is given by

(3-11) $\quad X_d = \sqrt{\dfrac{2\epsilon_s\epsilon_o\phi_s}{qN_A}}$

Equation (3-9) now reduces to

(3-12) $\quad V_G = \dfrac{-Q_B}{C_o} + \phi_s = \dfrac{\sqrt{2\epsilon_s\epsilon_o qN_A\phi_s}}{C_o} + \phi_s$

At inversion condition, $\phi_s = 2\phi_f$; thus the threshold voltage for an n-channel device is

(3-13) $\quad V_{Tn} = \dfrac{\sqrt{2\epsilon_s\epsilon_o qN_A(2\phi_f)}}{C_o} + 2\phi_f$

Under strong inversion, the silicon charge density now includes the inversion-layer charge Q_n:

(3-14) $\quad Q_s = Q_B + Q_n$

and

(3-15) $\quad V_G = \dfrac{-Q_B}{C_o} - \dfrac{Q_n}{C_o} + 2\phi_f$

$\qquad\quad = \dfrac{-Q_n}{C_o} + V_{Tn}$

Thus

(3-16) $\quad Q_n = -C_o(V_G - V_{Tn})$

For p-channel devices, the sign of Q_s changes from negative to positive because the depletion layer now contains positive donor charges, and ϕ_s and $2\phi_f$ are now negative:

(3-17) $\quad V_{Tp} = \dfrac{-\sqrt{2\epsilon_s\epsilon_o qN_D|2\phi_f|}}{C_o} + 2\phi_f$

and

(3-18) $\quad Q_p = -C_o(V_G - V_{Tp})$

3-3 NONIDEAL EFFECTS

There are several nonideal effects that perturb the ideal case described in Secs. 3-1 and 3-2. The nonzero ϕ_{MS} and nonzero Q_f will be specifically covered here, and an expression for these effects on V_T will be presented. Other nonideal effects are discussed in Chap. 4.

Nonzero ϕ_{MS}

Work function is the energy required to bring an electron from the Fermi level to the vacuum level. In a MOS structure, it is equivalent to the energy required to reach the oxide conduction band (see Fig. 3-5). The aluminum work function ϕ_M and silicon work function ϕ_S are not equal; thus when an aluminum-oxide and an oxide-semiconductor system are brought together as in Fig. 3-5, the work function difference will not be zero:

(3-19) $\quad \phi_{MS} = \phi_M - \phi_S \neq 0$

This results in a built-in charge as the Fermi levels on both sides of the oxide attempt to line up in Fig. 3-6, as they should in an equilibrium system. Figure 3-7 shows that an external voltage of an amount V_{FB} would need to be applied to bring the MOS system to the flatband condition. V_{FB} is called

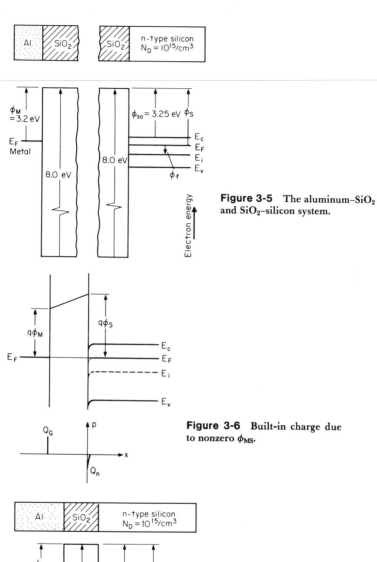

Figure 3-5 The aluminum–SiO$_2$ and SiO$_2$–silicon system.

Figure 3-6 Built-in charge due to nonzero ϕ_{MS}.

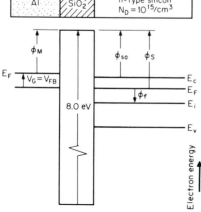

Figure 3-7 Application of V_{FB} to bring about flatband condition for n-type silicon.

the flatband voltage. Once V_{FB} is determined, by principle of superposition, all the previously developed expressions can still be utilized.

Since the metal-semiconductor work function difference ϕ_{MS} is measured from the gate Fermi level to the silicon Fermi level, its value would depend on the bulk semiconductor type, its doping concentration, and the gate material. Figure 3-7 shows the case for an n-type substrate. That is contrasted with the case for the p-type silicon in Fig. 3-8, while Fig. 3-9 shows the case for a polysilicon gate which is doped p-type. All situations are represented in Fig. 3-10 which plots the work function difference vs. doping concentration for various gate electrode materials. In the case of polysilicon gates, the polysilicon is assumed so heavily doped that it is degenerate, and

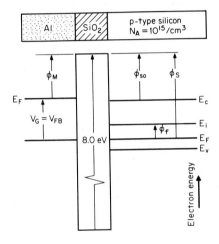

Figure 3-8 Application of V_{FB} to bring about flatband condition for p-type silicon.

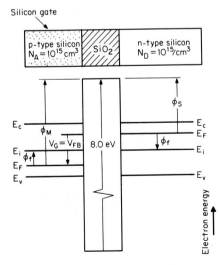

Figure 3-9 Application of V_{FB} to bring about flatband condition for p^+-doped polysilicon gate to n-silicon system.

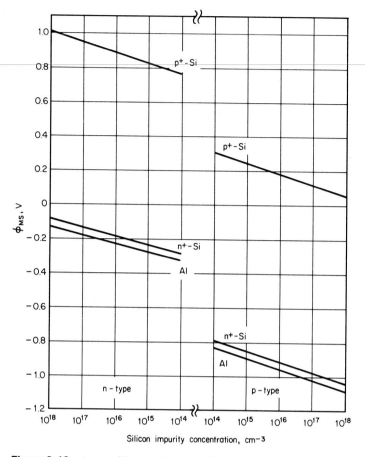

Figure 3-10 ϕ_{MS} **vs. silicon doping for various gate electrodes.**

the Fermi level of the gate electrode is pegged at either one of the two band edges.

Nonzero Fixed Oxide Charge $Q_f(Q_{ss})$

The next most important nonideal perturbation of the MOS system is the presence of some positive charge at the oxide-semiconductor interface. This is brought about by the oxidation process itself and is the result of dangling silicon bonds remaining after oxidation. It resides just slightly into the oxide side of the oxide-silicon interface and is relatively independent of oxide thickness, doping type, or doping concentration. The surface charge density is given as Q_f, in coulombs per square centimeter. Another symbol widely

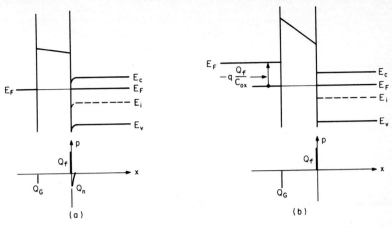

Figure 3-11 Band diagrams showing (*a*) band bending due to Q_f, even with no gate voltage applied (gate grounded), and (*b*) flatband voltage needed to restore flatband condition.

used is Q_{ss}. With the gate grounded, Q_f^* will also cause band bending (Fig. 3-11*a*). An external voltage needs to be applied to restore the flatband condition (Fig. 3-11*b*). This external voltage is another additive component to the flatband voltage. The required voltage to restore flatband is

$$(3\text{-}20) \quad V_{FB} = \frac{-Q_f}{C_o}$$

Since Q_f is always positive, its flatband voltage is negative.

Q_f is strongly affected by oxidation conditions, particularly the final oxidation temperature and ambient. This relationship is presented by the well-known Q_f triangle[4] of Fig. 3-12. At low temperature, the oxide formation is limited by the oxidation process itself; there is an abundance of dangling Si bonds waiting to combine with O_2, hence Q_f is high. At high temperatures, the oxide-forming process is limited by the diffusion rate of oxygen through the oxide. The oxidation process is principally complete, and Q_f is low. In any case, with subsequent annealing in inert gases such as nitrogen or argon, any excess Si bond is allowed time to complete the oxidation, and again Q_f is low. Figure 3-13 shows how annealing takes place as a function of time and temperature. That figure would indicate that annealing in nitrogen at about 1000°C, and for at least 20 min, would be optimum. Figure 3-14 shows an example of an actual gate oxidation specification that includes a 1-h anneal cycle in 1050°C argon for reducing Q_f.

---

*This book follows the recently established IEEE/Electrochemical Society standardized terminology for oxide charges associated with the thermally oxidized silicon system.[3]

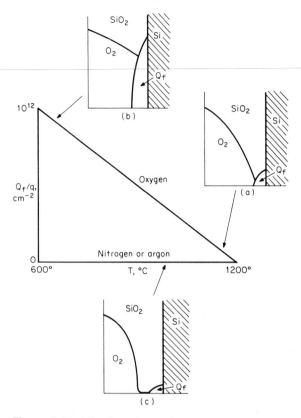

Figure 3-12 The dependence of Q_f on final oxidation temperature and ambient as represented by the Q_f triangle. Data are for (111) material, but the same general relationship holds for (100) material. Also shown are sketches of thermal oxide cross section. *(After Ref. 4. Reprinted with permission of the Electrochemical Society.)*

Q_f is also dependent on crystal orientation. Table 3-1 shows that the amount of Q_f is weakly correlated to the available bonds per unit area. In that respect, (100) material consistently has lower Q_f than (111) and hence is almost universally preferred in MOS work.

Taking Nonideal Effects into Account

All the nonideal effects can be added by the principle of superposition. In particular, the voltage loop equation, Eq. (3-12), can now be written as

(3-21) $$V_G = V_{FB} + \phi_s - \frac{Q_B}{C_o}$$

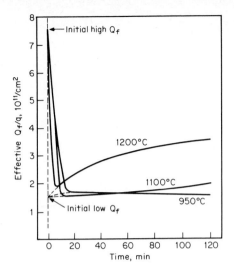

Figure 3-13 Reduction of Q_f vs. annealing time and annealing temperature. [0.12 μm SiO_2–Si (111) nitrogen anneal] *(After Ref. 4. Reprinted with permission of the Electrochemical Society.)*

Cycle:
    5 min slow push in O_2 at 800°C
    20 min ramp in O_2 from 800°C to 1050°C
    30 min dry O_2 at 1050°C
    21.5 min O_2/6% HCl—Adjust time for t_{ox} control
    60 min dry Ar at 1050°C
    60 min ramp down in Ar from 1050°C to 800°C
    5 min slow pull in Ar at 800°C
    10 min cool in ampoule with Ar

T_{ox}: 900 Å ± 65 Å

CV shift < 0.3 V

Figure 3-14 Use of argon anneal to reduce Q_f.

TABLE 3-1 Typical Q_f Values

Orientation	Available bonds per square centimeter	Typical Q_f/q
(111)	11.8×10^{14}	5×10^{11} cm^{-2}
(110)	9.6×10^{14}	2×10^{11} cm^{-2}
(100)	6.8×10^{14}	0.9×10^{11} cm^{-2}

At inversion,

$$(3\text{-}22) \quad V_G = V_T = \phi_{MS} - \frac{Q_f}{C_o} + 2\phi_f - \frac{Q_B}{C_o}$$

The first two terms are the voltage required to establish flatband, and the last two are the voltage required to bend the bands through $2\phi_f$.

ϕ_{MS} and Q_f are the two major nonideal effects of a MOS system. There are other effects, and they will be discussed in conjunction with CV plots in Chap. 4. CV plots allow one to identify these nonideal effects quite readily.

REFERENCES

1. P. Richman, *MOS Field-Effect Transistors and Integrated Circuits,* Chap. 2, Wiley, New York, 1973.

2. S. M. Sze, *Physics of Semiconductor Devices,* 2d ed., Chap. 8, Wiley, New York, 1981.

3. B. Deal, "Standardized terminology for oxide charges associated with thermally oxidized silicon," *IEEE Trans. Elect. Dev.,* vol. ED-27, no. 3, pp. 606–608, March 1980.

4. B. Deal, "The current understanding of charges in the thermally oxidized silicon structure," *J. Electrochem. Soc.,* vol. 121, no. 6, pp. 198C–205C, June 1974.

PROBLEMS

1. Draw the charge distribution under (**a**) accumulation, (**b**) depletion, (**c**) strong inversion, similar to Fig. 3-4, but for n-type substrate. Indicate the polarity of the applied voltage.

2. The charge distribution for a p-type substrate biased near inversion is represented in Fig. 3-15. By examining the definition for ϕ_f and the definition of n in terms of E_i and E_F, draw the size of Q_n *right at* inversion, i.e., when $\phi_s = 2\phi_f$.

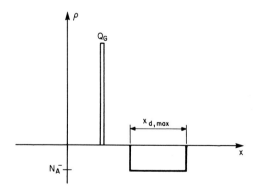

Figure 3-15   Problem 2.

3. Calculate the threshold voltage for an aluminum gate MOS structure with $N_A = 5 \times 10^{15}$ cm^{-3}, and $Q_f/q = 1 \times 10^{11}$ charges per square centimeter. Oxide thickness = 1000 Å. Repeat for the case where the substrate is n type with $N_D = 5 \times 10^{15}$ cm^{-3}.

4. In the LOCOS process (see Chap. 7), n^+ doped polysilicon is often used as the interconnection between MOS devices. These devices are separated by a thick field oxide. This polysilicon forms the gate of a MOS device with the thicker field oxide as its gate oxide. The process is designed such that its turn-on voltage, the field threshold, has to be greater than the highest voltage applied to the polysilicon, i.e., the power supply voltage. Calculate the field threshold voltage for a field oxide thickness of 8000 Å, p-type substrate implanted to a concentration of 5×10^{16} cm^{-3}. Repeat the calculation for an aluminum interconnect with an *additional* 4000 Å of glass deposited in between. See Fig. 3-16 for a cross section.

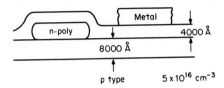

Figure 3-16   **Problem 4.**

5. Take the derivative of V_T with respect to temperature, and come up with an expression for the change in V_T with respect to temperature. Do it for both the p-type and the n-type substrates. If ϕ_{MS} and Q_f do not change with the temperature, while n_i doubles roughly every 8°C, what general rule can you formulate about the effect of temperature on V_T?

6. What happens to the threshold of a device if its heavily n^+-doped polysilicon gate is changed to heavily p^+-doped polysilicon? Give a quantitative answer.

4

CV Plots

4-1 THE IMPORTANCE OF *CV* PLOTS

The single most important measurement in MOS work is a plot of the capacitance of a MOS structure as a function of its dc gate bias. This is known as a *CV* (capacitance-voltage) plot.[1] Its importance lies in the wealth of information that can be gathered from a single plot and from the fact that it can be performed on a relatively simple structure, permitting, for example, in-process quality control monitoring.

A typical test structure and setup is shown in Fig. 4-1. The test structure consists of a gate electrode (often metal and usually in a circular dot to minimize fringe effects) over a gate oxide. This forms a simple MOS capacitor. The capacitance meter measures the capacitance of the structure while allowing the application of dc bias. By allowing the dc bias to change and plotting the capacitance as a function of the dc bias, the *CV* plot of Fig. 4-2 results.

4-2 HIGH-FREQUENCY *CV* PLOTS

Capacitance meters measure capacitance by applying a small, typically 50 mV, ac voltage on top of the dc bias and measuring the reactive component of the resulting current. When a relatively high signal frequency is used, commonly 1 MHz, the resulting *CV* plot is characteristically given by Fig. 4-2. This is the high-frequency *CV* plot and is by far the most commonly used type of *CV* plot.

As indicated in Fig. 4-2, with a negative gate bias, *p*-type silicon is in accumulation, and one measures simply the capacitance of a parallel plate capacitor with the oxide as the dielectric. At a gate voltage that is more

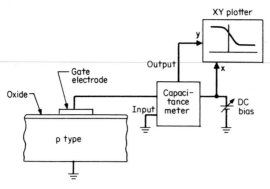

Figure 4-1   *CV* plot test structure and setup.

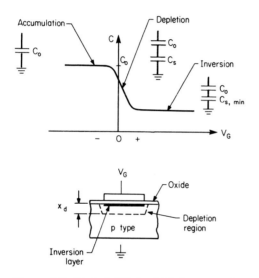

Figure 4-2   Typical *CV* plot (high-frequency) for
p-type substrate.

positive than the flatband voltage V_{FB}, a depletion layer is formed in the
semiconductor. This creates a capacitor in series with the oxide capacitor
and produces a drop in total capacitance. When the dc gate voltage reaches
and exceeds the threshold voltage V_T, a layer of inversion charge is formed.
Then if the dc gate voltage is slowly increased further, the inversion layer
increases its charge to balance the gate and the depletion layer does not
widen further. Since the depletion-layer width has reached a maximum, the
total ac capacitance is pegged at its minimum.

4-3 LOW-FREQUENCY *CV* PLOTS

When the test frequency of the capacitance meter is very low, the *CV* plot looks more like Fig. 4-3. Just as in the high-frequency case, the capacitance starts out at a maximum at accumulation and drops in depletion. However, when the inversion condition is reached, the inversion layer can actually change its charge at the same rate as the ac signal. Since the inversion layer physically resides at the oxide-semiconductor interface, one is again measuring the oxide capacitance, and the total capacitance increases again to the maximum. It should be pointed out that the threshold voltage does not correspond to the capacitance minimum but is slightly beyond that. The transition from a high-frequency plot to a low-frequency one is distinctly displayed in Fig. 4-4.

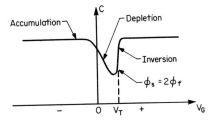

Figure 4-3 Typical low-frequency *CV* plot.

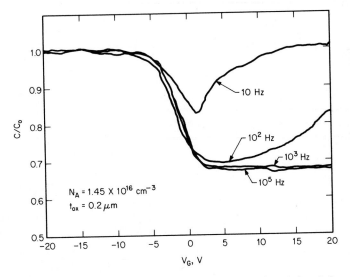

Figure 4-4 Effect of measurement frequency on *CV* plots. (*After A. S. Grove et al., J. Appl. Phys., vol. 35, p. 2458, 1964.*)

   The difference between high- and low-frequency CV plotting is further elucidated by Fig. 4-5. It shows the fluctuation in charge distribution in the inversion condition in response to the fluctuation of the ac gate signal. Figure 4-5a is the high-frequency case. It shows that at inversion, there is indeed a sheet layer of charge at the oxide-semiconductor interface, but the ac fluctuation of the gate charge occurs at a high enough rate that only the widening and narrowing of the depletion layer can follow the charge fluctuation. This widening and narrowing takes place at the incredibly fast dielectric relaxation rate, which is on the order of 10^{-14} s. In Fig. 4-5b for the low-frequency case, the gate charge fluctuates slowly enough that the inversion charge can follow the variation directly. The frequency needs to be in the 10- to 100-Hz range before this can happen. In both cases, note that the depletion layer remains at its maximum width.

   The measurement frequency that causes the MOS capacitor to exhibit a high- rather than low-freqeuncy CV plot depends on the device itself. This frequency must be high enough such that the thermal generation of carriers

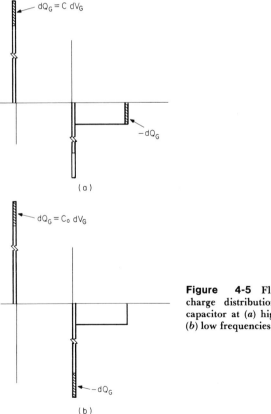

(a)

(b)

Figure 4-5 Fluctuation in charge distribution in a MOS capacitor at (a) high frequencies, (b) low frequencies.

cannot follow the gate voltage variation. A commonly used frequency is 1 MHz. A low frequency is one for which the inversion-layer charge can be increased or decreased at the same rate as the small ac signal on the gate. Under this definition, the *CV* plot of the gate of a MOS transistor exhibits a *low-frequency* plot if its source and/or drain are grounded. In that condition, the inversion charge can readily flow in and out from under the gate via the source and drain. A low-frequency plot always results even at a high measurement frequency. The transistor gate is rarely *CV*-plotted because its small size causes the measurement to be swamped by parasitic capacitances; but it is technically feasible, and by leaving the source and drain floating, the high-frequency plot can be obtained.

4-4 EQUATIONS FOR *CV* PLOTS

The high-frequency capacitance-vs.-voltage equation for a MOS capacitor will be derived in this section. All expressions for capacitances are per unit area. Furthermore, the semiconductor is assumed uniformly doped.

When the MOS capacitor is in accumulation, the measured total capacitance is

(4-1) $C = C_o$ (Accumulation)

where C_o is the oxide capacitance. In depletion, the capacitance drops with increasing gate voltage. To find the precise relationship, first recall that

(3-12) $V_G = \dfrac{-Q_B}{C_o} + \phi_s = \dfrac{\sqrt{2\epsilon_s\epsilon_o q N_A \phi_s}}{C_o} + \phi_s$

Then

(4-2) $\dfrac{dV_G}{dQ_B} = -\dfrac{1}{C_o} + \dfrac{d\phi_s}{dQ_B}$

Since the incremental gate charge has to be balanced by an opposite incremental depletion-layer charge, then $dQ_B = -dQ_G$. Also, the depletion-layer capacitance $1/C_s = -d\phi_s/dQ_B$. Thus

(4-3) $\dfrac{1}{C} = \dfrac{dV_G}{dQ_G} = \dfrac{1}{C_o} + \dfrac{1}{C_s}$

Equation (4-3) is a restatement of the fact that the total capacitance is the series combination of the oxide capacitance and depletion-layer capacitance. Another form of the same equation is

(4-4) $\dfrac{C}{C_o} = \dfrac{1}{1 + C_o/C_s}$

All that remains is to express C_o/C_s as a function of V_G. One refers again to the voltage equation

(3-12) $$V_G = \frac{-Q_B}{C_o} + \phi_s$$

Expressions for Q_B and ϕ_s are similar to those of Eqs. (2-35) and (2-37):

(4-5) $$V_G = \frac{qN_A x_d}{C_o} + \frac{qN_A x_d^2}{2\epsilon_s\epsilon_o}$$

Since $C_s = \epsilon_s\epsilon_o/x_d$, then

(4-6) $$V_G = \frac{qN_A\epsilon_s\epsilon_o}{C_o C_s} + \frac{qN_A\epsilon_s\epsilon_o}{2C_s^2}$$

By defining a constant

(4-7) $$\frac{1}{V_o} = \frac{2C_o^2}{qN_A\epsilon_s\epsilon_o}$$

and multiplying Eq. (4-6) by it and rearranging terms,

(4-8) $$\left(\frac{C_o}{C_s}\right)^2 + \frac{2C_o}{C_s} - \frac{V_G}{V_o} = 0$$

Solving Eq. (4-8) for C_o/C_s and taking the meaningful root, we obtain

(4-9) $$\frac{C_o}{C_s} = -1 + \sqrt{1 + \frac{V_G}{V_o}}$$

Substituting Eq. (4-9) into Eq. (4-4),

(4-10) $$\frac{C}{C_o} = \left(\sqrt{1 + \frac{V_G}{V_o}}\right)^{-1} = \left(\sqrt{1 + \frac{2V_G C_o^2}{qN_A\epsilon_s\epsilon_o}}\right)^{-1} \qquad \text{(Depletion)}$$

The above derivation assumes that $V_{FB} = 0$. If that is not the case, replace V_G in Eq. (4-10) with $(V_G - V_{FB})$.

Once inversion is reached, the capacitance stays at a constant minimum value of

(4-11) $$\frac{C_{min}}{C_o} = \left(\sqrt{1 + \frac{V_T}{V_o}}\right)^{-1} \qquad \text{(Inversion)}$$

where V_T is the threshold voltage. Figure 4-6 shows the comparison between a more exact analysis and Eqs. (4-1), (4-10), and (4-11). The agreement is good except for more rounding of the curve for the more exact analysis, particularly at the flatband condition ($V_{FB} = 0$ for this ideal case).

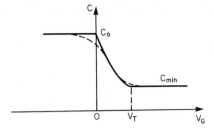

Figure 4-6 Comparison between more exact analysis and Eqs. (4-1), (4-10), and (4-11) in describing *CV* plots. The continuous line represents simple theory; the dashed line represents exact analysis.

More exact analysis[2,3] shows the depletion-layer capacitance at flatband not to be infinite, even though depletion width shrinks to zero, but rather

$$(4\text{-}12) \quad C_{sFB} = \frac{\epsilon_s \epsilon_o}{L_D}$$

where L_D, the extrinsic Debye length, is

$$(4\text{-}13) \quad L_D = \sqrt{\frac{\epsilon_s \epsilon_o k T}{q^2 N_A}}$$

Thus

$$(4\text{-}14) \quad C_{sFB} = \sqrt{\frac{q \epsilon_s \epsilon_o N_A}{kT/q}}$$

At flatband, the total capacitance normalized to the maximum capacitance is thus

$$(4\text{-}15) \quad \frac{C_{FB}}{C_o} = \frac{C_{sFB}}{C_o + C_{sFB}}$$

Figure 4-7 shows the normalized flatband capacitance as a function of oxide thickness for various doping concentrations.[3] It is useful in determining the flatband voltage shift, which is discussed later in Sec. 4-7.

4-5 NORMALIZATION OF THE IDEAL *CV* PLOT

The *CV* equations derived in the previous section indicate that the *CV* plot can be normalized to its maximum value C_o without loss of generality. This removes the need to know the area of the MOS capacitor, and other information can be obtained more quickly. One such normalized curve is shown in Fig. 4-8.

From Eq. (4-10) can be deduced the fact that in depletion, at any applied voltage V_G, the capacitance will be higher (i.e., *CV* plot flatter) for higher N_A and/or larger t_{ox}. This fact is displayed in Fig. 4-9a and b. The threshold voltage, the voltage at which C_{min} is reached, is larger (for *p*-channel

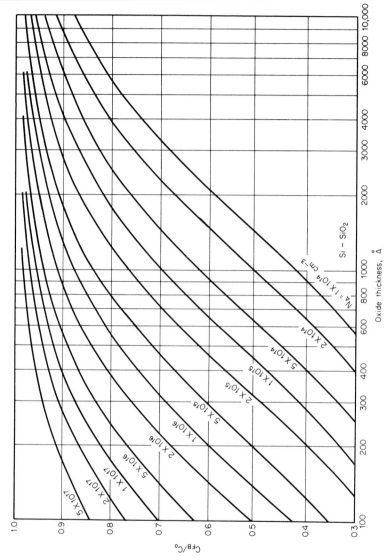

Figure 4-7 Normalized flatband capacitance as a function of oxide thickness for various doping concentrations. (*After Ref. 3.*)

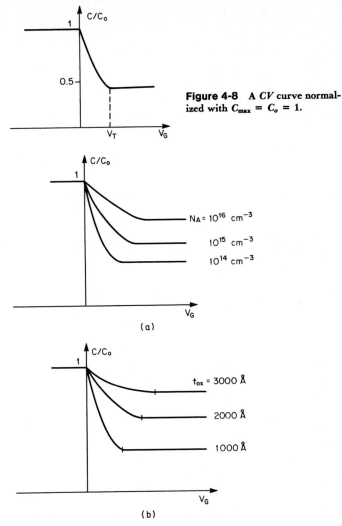

Figure 4-8 A *CV* curve normalized with $C_{max} = C_o = 1$.

Figure 4-9 Normalized *CV* curve for (*a*) fixed oxide thickness and variable doping concentration, (*b*) fixed doping concentration and variable oxide thickness.

devices, larger in magnitude) for higher N_A and t_{ox}. A similar conclusion would have been drawn based on the expression for threshold voltage developed earlier, Eqs. (3-13) and (3-17).

4-6 DEEP DEPLETION

In generating the *CV* plot, if the MOS capacitor is swept from accumulation toward inversion at a relatively fast rate (about 10 V/s and higher), one often finds that instead of flattening out at the inversion value, the capacitance continues to drop, following the depletion curve (see Fig. 4-10). This drop results when the minority carriers cannot be generated fast enough to balance the changing dc gate voltage. When that situation exists, the depletion layer has to widen to balance the increased gate charge. The depletion layer widens past its "maximum" value (when $\phi_s = 2\phi_f$), and the capacitance drops below its C_{min} value.

If minority carrier lifetime is extremely short, such as with gold-doped material, the deep-depletion characteristics may not be observed with normal *CV* plotting procedures.

The deep depletion is a nonequilibrium condition. If the *CV* plot is stopped at any point in the deep-depletion condition, thermal generation of carriers will bring the capacitance value back to C_{min} within a few seconds. Shining light on the device will accomplish the same effect immediately. This nonequilibrium deep depletion is precisely the condition that charge-coupled devices (CCDs) operate under. By continuous clocking of a series of closely spaced MOS capacitors, packets of inversion-layer charge serving as the signal can be propagated along a shift register. Since the device is in deep depletion, minority carriers are constantly being generated and added to the signal charge, thus corrupting it to some degree.

4-7 DEVIATIONS FROM THE IDEAL *CV* PLOT

The usefulness of a *CV* plot lies in the ability to extract so much information about various nonideal mechanisms from very simple structures. Examples of some of them and their effects on the CV plot are discussed below.

Nonzero V_{FB}

A flatband voltage that is different from zero will cause a horizontal parallel shift in the *CV* plot. Figure 4-11 shows that a positive V_{FB} causes a shift to

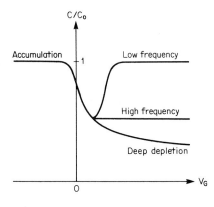

Figure 4-10 Deep-depletion *CV* plot, superimposed on low-frequency and high-frequency *CV* plots.

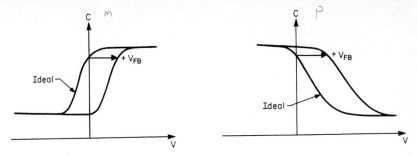

Figure 4-11 $V_{FB} > 0$ causes the *CV* plot to be shifted to the right.

the right on *p*-channel and *n*-channel devices alike. If V_{FB} is negative, the shift is to the left.

Nonzero ϕ_{MS}

Nonzero ϕ_{MS} also causes a parallel shift in the *CV* plot. Figure 4-12 is the result of a unique experiment[4] in which various metals were used as gate electrodes. The parallel shift caused by varying ϕ_{MS} is dramatic. While it is unlikely that modern MOS processing will use such gate materials, the same principle dictates that when one switches from a p^+-doped polysilicon gate to an n^+-doped polysilicon gate, the *CV* plot shifts by 1.1 V (which direction?). The flatband shift due to ϕ_{MS} is

(4-16) $V_{FB,shift} = \phi_{MS}$

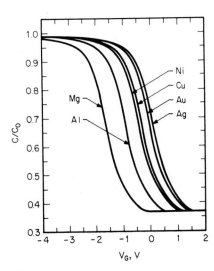

Figure 4-12 *CV* plot with several values of ϕ_{MS}. *(After Ref. 4. Reprinted with permission of Pergamon Press.)*

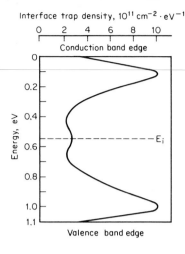

Figure 4-13 Typical surface state density as a function of energy level. (*After Ref. 11. Reprinted with permission of the Electrochemical Society.*)

Q_f (Fixed Oxide Charge)

The flatband voltage shift due to Q_f is given by

$$(4\text{-}17) \qquad V_{FB,shift} = \frac{-Q_f}{C_o}$$

where Q_f is in coulombs per square centimeter, and C_o is the oxide capacitance in farads per square centimeter. Equation (4-17) is also applicable to any sheet-layer charge that exists right at the oxide-semiconductor interface. Since Q_f is always a positive charge, the flatband shift caused by it is always negative. Typical values of the V_{FB} shift from Q_f are -0.5 V or less. One other symbol in common use is Q_{ss}.*

D_{it} (Interface Traps)

Interface traps are electronic states that reside at the oxide-semiconductor interface. In fact, they are the results of dangling bonds. However, unlike Q_f, rather than having a fixed charge, interface traps are charged or discharged (positive or negative, depending on whether the states are donor-like or acceptor-like) with changes in band bending.

Interface traps are electronic states situated within the band gap. Of course it is very possible that they also exist within the conduction and valence bands but are indistinguishable by measurement from the large density of band states. Rather than being a single discrete energy level, interface traps are a continuum of energy levels; thus they are quantified as interface

---

*This book follows the IEEE/Electrochemical Society standard terminology for oxide charges associated with the thermally oxidized silicon system.[10]

trap *density* D_{it} with a unit of number of states/(cm²·eV). Good modern MOS processing can produce devices with a D_{it} of $\approx 10^{10}$ cm⁻²·eV. Figure 4-13 shows the typical distribution of surface state as a function of energy level. It forms two high peaks near the band edges and a minimum near midgap. The midgap levels are the ones primarily involved with the carrier generation *at the surface* of a MOS capacitor in depletion.

To visualize the effect of interface traps on the *CV* plot, refer to Fig. 4-14, which shows *p*-type silicon with donor-like states uniformly distributed at the surface. In Fig. 4-14*a*, with the silicon in accumulation, all the interface traps would be positively charged. This charge results from the fact that states below the Fermi level are considered filled with electrons (donors remain neutral), while states above the Fermi level are considered emptied of electrons (donors become charged). As the device moves toward inversion, the amount of positive charge at the surface reduces to a minimum. The resulting *CV* plot (Fig. 4-15) shows a maximum negative shift in accumulation, gradually changing to almost no shift near inversion. The net result is a *CV* plot with a gentler slope, a distinct indicator of high surface state density. Since the amount of shift is variable, a common point of measure-

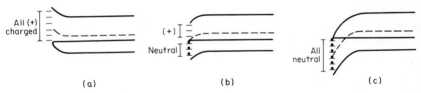

Figure 4-14 Change in net charge of donor-like surface states at the interface due to band bending.

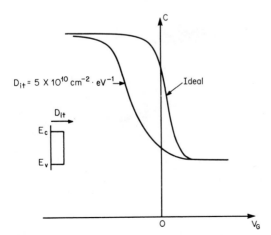

Figure 4-15 *CV* plot shift due to uniform donor-like surface states.

ment needs to be specified when talking about CV shifts. The flatband condition is one such measurement point, although one can see that it does not completely quantify the problem.

Interface traps have been referred to by others as surface states and fast surface states; other symbols sometimes used are N_{ss}, N_{FS}, or N_{st}.

Mobile Ions

A trace amount of sodium ions, Na^+, in the gate oxide would cause a CV shift. However, the mere fact that a CV plot would be shifted from the ideal is not as big a concern as the fact that (1) the amount of ions is variable from device to device, and that (2) sodium ions move around inside the oxide, causing long-term instability in the device threshold.

The flatband shift caused by an arbitrary charge density distribution, such as shown in Fig. 4-16, is

$$(4\text{-}18) \qquad V_{FB,shift} = \frac{-1}{\epsilon_{ox}\epsilon_o} \int_0^{t_{ox}} x\rho(x)\ dx$$

The above equation states that if one applies a negative gate bias and pushes all the positive sodium ions in the oxide to $x = 0$, i.e., to the metal-oxide interface, then the flatband shift would be zero. This is depicted in Fig. 4-17a. When a positive gate voltage is applied, such that all ions are at $x = t_{ox}$ as in Fig. 4-17b, then the flatband shift is maximum.

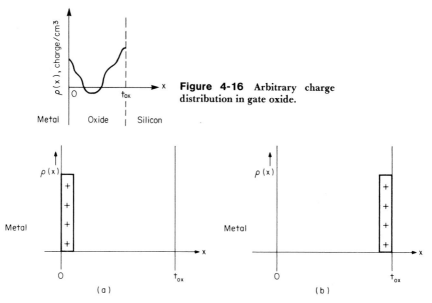

Figure 4-16 Arbitrary charge distribution in gate oxide.

Figure 4-17 (*a*) Flatband shift is zero when all gate ionic charges are at the metal-oxide interface; (*b*) flatband shift is maximum when all charges are at oxide-semiconductor interface.

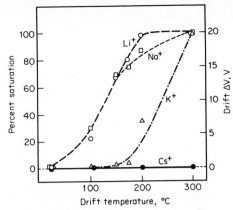

Figure 4-18 Comparison of drift rate for various alkali ions through 0.2-μm oxide. Drift field = +50 V/μm; drift time = 2 min; 46 ppm alkali chloride. (*After Ref. 11. Reprinted with permission of the Electrochemical Society.*)

Sodium is the most common ionic contaminant in MOS processing. It leaches from glass containers into chemicals used in processing and from direct or indirect contact with body salts. Other alkali ions such as lithium and potassium are possible contamination sources too. Figure 4-18 shows the drift rate that results when a bias is applied across a 2000-Å oxide at elevated temperatures. The temperature at which a significant voltage shift initiates can be used as a means to distinguish between potassium and the other ion species. The actual procedure is to apply a bias across the MOS capacitor and slowly raise the temperature. At a high enough temperature, the ions will gain enough mobility to move in response to the field. This movement in gate charge is detected as a small but distinct gate current.

The unique property of the mobile ions to move under high field and high temperature has led to the development of the bias-heat stress technique of detecting mobile ions. The steps are as follows:

1. Plot *CV* at room temperature. This typically results in curve 1 of Fig. 4-19 for a *p*-substrate capacitor.

2. With +30 V applied to the gate, heat to 250°C for 3 min. Figure 4-18 shows that all available sodium ions would have moved to the oxide-semiconductor interface. Cool to room temperature under bias and do a *CV* plot. One should see a large shift to the left, such as curve 2 in Fig. 4-19.

3. Then, finally, with −30 V applied to the gate, heat to 250°C for 3 min. Cool to room temperature and do a *CV*. One should see a large shift to the right, such as curve 3, even passing the initial curve 1 where the ions are distributed throughout the gate oxide.

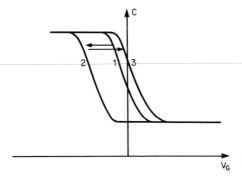

Figure 4-19 Shifts in *CV* plot with bias-heat stress. Curve 1, initial; curve 2, after positive bias and high temperature (plotted after cooling to room temperature); curve 3, after negative bias and high temperature (plotted after cooling).

The *CV* plot of Fig. 4-19 is often observed for MOS capacitors that are definitely contaminated with sodium. But this is a rare occurrence in modern MOS processing. More commonly, curves 1, 2, and 3 differ by no more than 0.1 V. It is still good practice to do a routine *CV* plot as one of several monitors of processing quality, because unexpected sources of contaminants are plentiful, from a bad batch of one chemical to leaky vacuum systems contaminated with diffusion pump oil. Fabrication areas that are converting to MOS from other semiconductor technologies should be particularly sensitive about cleanliness and purity. The discipline for cleanliness necessary for modern MOS work sometimes escapes the uninitiated.

A common variation[1] from Fig. 4-19 is that curve 3 ends up in between curves 1 and 2, though often quite close to the original curve 1. This is explained in part by the additional effect of charge trapping. In unannealed or improperly annealed oxides, under strong bias, charges can be injected by tunneling from the bulk silicon and remain trapped in the first 100 Å of the oxide. A positive bias would inject electrons and a negative bias would inject holes. On devices where this effect is dominant, a positive bias will produce a positive *CV* shift and a negative bias will produce a negative *CV* shift, completely opposite to the direction obtained for movement of ion in the oxide. Hydrogen anneals have been very successful in reducing this effect.

Radiation Effects

Exposing MOS devices to high-energy ionizing radiation can lead to trapping of holes in the oxide.[5] The radiation ionizes the insulator by generating electron-hole pairs in it. The electrons readily pass through the oxide-semiconductor interface while the holes are captured by traps. This effect is accentuated when a positive gate voltage is applied during irradiation. Even after the radiation source is removed, a layer of positive charge remains at the oxide-semiconductor interface. This causes a shift of the *CV* plot to the left. A typical change in threshold voltage as a function of radiation dose, in rad (into silicon), is shown in Fig. 4-20. This flatband shift can be annealed out at low (300°C) temperature. Such a treatment is helpful if radiation occurs during normal device processing such as electron-beam metallization, ion implantation, or sputtering, but the treatment is obviously unavailable if radiation occurs during a space flight or from nuclear explosion.

Radiation hardening of MOS circuits can be achieved by growing the gate oxide in such a manner that it has the property of allowing the trapped holes to leak off readily. Another solution is surprisingly simple and is suggested by Fig. 4-20, that is, design the process such that the n-channel devices have a higher threshold. Radiation will then bring it back to the desired value. A related radiation phenomenon is that during a high *rate* of radiation, large currents are generated in junction depletion regions that can cause logic upset. One solution in CMOS circuits is to design p-channel and n-channel transistors to have the same drain areas such that the generated currents will cancel each other.

Radiation can also generate new interface traps and create electron traps near the interface. These will cause distortion and hysteresis in the CV plot. However, these effects are not as severe a detriment as the trapping of holes.

As a review, the four types of charges associated with a thermally grown oxide-semiconductor interface are shown in Fig. 4-21. Q_f and D_{it} are located right at the interface, while mobile ions are distributed throughout the oxide. Ionized traps are in the first 100 Å of the oxide.

Impurity Profile

The CV plot can be used to probe the impurity profile of the substrate. Differentiation of the normalized depletion capacitance of Eq. (4-10) results in

$$(4\text{-}19) \quad \frac{d(C/C_o)}{dV_G} = -\frac{1}{2}\left(\frac{C}{C_o}\right)^3 \frac{1}{V_o} = -\left(\frac{C}{C_o}\right)^3 \frac{\epsilon_{ox}^2 \epsilon_o}{\epsilon_s q N_A t_{ox}^2}$$

This states that for any fixed value of the normalized capacitance C/C_o, the slope of the CV plot is inversely proportional to the product $N_A t_{ox}^2$.

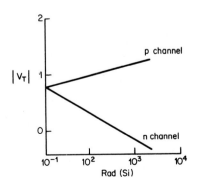

Figure 4-20 Shifts in threshold voltage for p- and n-channel devices as a function of radiation dose, in rad (into silicon).

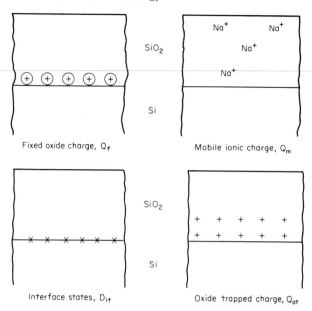

Figure 4-21   **Four types of charges associated with the thermally grown oxide-semiconductor interface.** (*After Ref. 11. Reprinted with permission of the Electrochemical Society.*)

At any point on the *CV* plot the capacitance meter is probing only the doping concentration at the edge of the depletion layer. If Eq. (4-19) is used to calculate N_A corresponding to various points on the *CV* plot, a doping profile of the silicon substrate can be obtained. Equation (4-19) is valid only for a *CV* plot that is free of the distortion introduced by interface traps. Recall that interface traps change the slope of *CV* plots.

In interpreting the doping concentration obtained from *CV* plots, one should be aware of the redistribution of impurities that can occur during

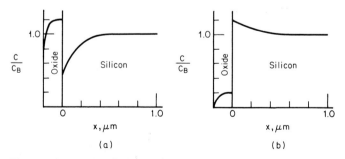

Figure 4-22   **Redistribution of impurities after oxidation for (a) boron, where oxide *takes up* impurity ($m < 1$), and (b) phosphorus, where oxide *rejects* impurity ($m > 1$).** (*After A. S. Grove et al., J. Appl. Phys., vol. 35, no. 9, p. 2695, September 1964.*)

thermal oxidation. (See Sec. 8-2.) When wafers of uniform doping concentration are thermally oxidized, the oxide layer will either take up or reject the impurity as it grows, creating a nonuniformity in doping concentration. Figure 4-22 shows that boron tends to be depleted near the surface, while phosphorus tends to "snowplow" and pile up. This effect is most severe, that is, the change in concentration is most pronounced for wet oxidation and for oxidation at lower temperatures.

4-8 PRACTICAL CONSIDERATIONS IN DOING *CV* PLOTS

Light and Temperature

The device under test in a *CV* plot should not be exposed to light. Figure 4-23 shows that under increasingly stronger illumination, the inversion capacitance increases until the low-frequency behavior is reached. This results from the generation of electron-hole pairs by photons which contribute to the inversion layer in balancing the gate charge. At any given gate voltage, the band bending needed to sustain a given inversion charge density is reduced. This shrinks the depletion-layer width and increases the inversion capacitance.

The effect of temperature on the device under test is similar to that of light (Fig. 4-24). For this reason, all *CV* plots should be performed only with the device at room temperature. Otherwise, the capacitance readings, $C_{\min}$ in particular, would be incorrect and would need a correction factor for temperature.

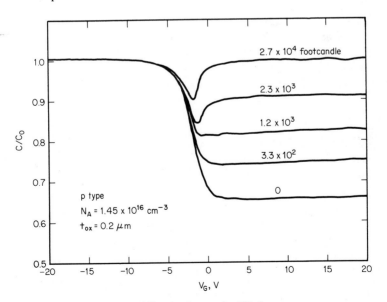

Figure 4-23 The effect of illumination on the *CV* plot.

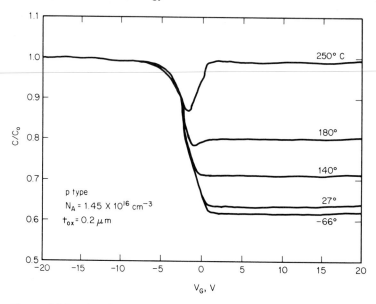

Figure 4-24 The effect of temperature on the *CV* plot.

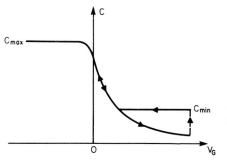

Figure 4-25 Performing the *CV* plot in both directions.

Sweep Rate

Sweeping from accumulation to depletion while using convenient sweep rates (1 to 10 V/s) and typical materials will usually result in the deep-depletion curve. By holding the maximum voltage at the inversion region, the capacitance will slowly rise to the inversion value C_{min} (Fig. 4-25). Illuminating the device at this point will speed up the process. Once C_{min} is reached, plotting from inversion to accumulation will trace out the familiar high-frequency *CV* plot. With stable devices fabricated in a clean environment, *CV* plots in the two directions will lie right on top of each other in the accumulation and depletion regions.

Good Contact to Substrate

If contact is made to the substrate through the back of the wafer, then the back of the wafer cannot have any residual oxide (indicated by presence of color). Any oxide introduces a capacitance in series with the device, reducing all capacitance readings. The oxide can be removed by swabbing with hydrofluoric acid. Furthermore, the wafer should be tightly clamped to the chuck by turning on the vacuum to maintain good mechanical contact.

A second problem dealing with contact to substrate frequently arises in testing capacitors located inside *p*-tubs in CMOS circuits. If the back of the wafer is used as the substrate connection instead of the *p*-tub contact, the *p*-tub capacitance will be introduced in series with the device, again reducing all capacitance readings. An example of this is shown in Fig. 4-26.

Moisture on Wafer Surface

For MOS capacitors that consist only of metal dots on a wafer with gate oxide, the erroneous plot of Fig. 4-27 is sometimes obtained. Through accumulation and depletion, the plot is normal. But when a voltage greater than

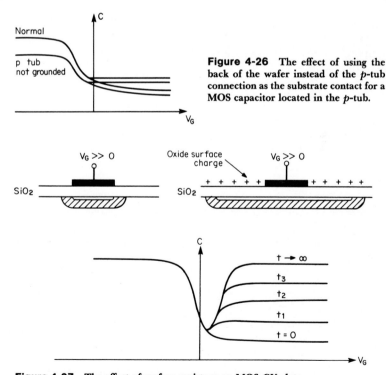

Figure 4-26 The effect of using the back of the wafer instead of the *p*-tub connection as the substrate contact for a MOS capacitor located in the *p*-tub.

Figure 4-27 The effect of surface moisture on MOS *CV* plots.

threshold is applied, the moisture on the oxide surface conducts the gate voltage to the area surrounding the metal dot, inverting it too. The depletion-layer capacitance is now higher because of the increased area. This effect becomes a problem only in the inversion region since only in the presence of the inversion layer can the larger depletion-layer area be coupled to affect the measurement.

The solution is to perform the *CV* plot within a few hours of fabrication of the metal dots, or to put devices through a stabilization bake of 250°C for 1 h (unbiased) in N_2 before measurement. More complicated solutions involve surrounding the MOS dot with field oxide, guard ring diffusions (same doping type as the substrate, also called channelstop), or a metal ring that is separately always biased in accumulation.

Miscellaneous

With most capacitance meters, a gate oxide short will result in negative capacitance readings. This is in contrast to the theoretical expectation that a short in a capacitor will result in infinite capacitance.

An easily overlooked precaution is to null out the stray capacitance of the cables in the setup before starting *CV* plots.

Finally, patterning the metal dots with photoresist followed by an etching step will give a better-defined metal pattern and more consistent results than forming the dots by evaporating metal through a shadow mask.

4-9 *CV* ANALYSIS PROGRAM

A program has been written to extract various information from a *CV* plot. The program is listed in the appendix at the end of this chapter and is written in BASIC. It assumes uniform doping concentration. Figure 4-28 shows a typical run of the program. First it asks for documentation such as mask set, lot number, and wafer identification. The user response is shown underlined. The program then asks whether it is *p*- or *n*-channel type. This corresponds to *n*- or *p*-type substrate, respectively. It next asks whether the gate material is aluminum, n^+ poly, or p^+ poly. The appropriate input is either A, N, or P, correspondingly. The next inputs are C_{max}, C_{min}, and area, all separated by commas. C_{max} and C_{min} are in picofarads, and area in square mils. C_{min} should be the minimum capacitance taken from the flat portion of the high-frequency plot and not the lowest reading taken from the deep-depletion curve. The program then outputs the flatband capacitance. The voltage corresponding to this capacitance is the flatband voltage and should be read off the *CV* plot and entered as an input. At this point the program is ready to print all the outputs.

From the maximum capacitance and the MOS capacitor area, the oxide thickness can be obtained. From the maximum and minimum capacitance, and the oxide thickness, the program calculates the doping concentration, work function difference, depletion charge, and depletion-layer width. Given the flatband voltage, it goes on to calculate the threshold voltage. Any difference between the ideal and actual flatband voltage is attributed to a net interface charge density. It should be noted that from a single *CV* plot, one cannot ascertain whether the net interface charge, and its resultant flatband shift, is due to Q_f, D_{it}, mobile ions, radiation-induced charge, or any other charges. One would need other measurements such as bias-heat stress or other techniques to distinguish between the various types of charges.

The program is written in the Applesoft version of BASIC. The example of Fig. 4-28 was run on an Apple II Plus computer. While every effort has been made to write the program in a widely used form of BASIC, readers who plan to use other interpreters should go through the listing to verify that all statements are compatible.

This program is based on an earlier version written in FORTRAN by Arnie London and Linda Collins at Motorola, Inc.

4-10   MEASUREMENTS OF MINORITY CARRIER LIFETIME

The minority carrier lifetime is related to many device properties. In dynamic memory circuits, it limits the time between refresh cycles that update the data. In sample and hold circuits, it limits the time a charge can be held before the junction leakage current reduces its magnitude. In charge-coupled devices (CCDs), it limits the low-frequency operation because thermally generated "dark current" distorts by adding to the signal charge.

Techniques have been developed to measure the minority carrier lifetime using device structures and instruments quite similar to those for *CV* plots. The two most popular techniques will be discussed below.

Capacitance-vs.-Time (*CT*) Plot

The *CT* plot is often employed because it can use the same MOS capacitor as for the *CV* plot, and many *CV* plotting instruments[6] have the capability for doing *CT* plots already built-in.

The *CT* plot is initiated by pulsing the gate electrode of a MOS capacitor from accumulation toward inversion using the voltage step function depicted in Fig. 4-29a. Because of the rapid voltage change, the capacitor will go into deep depletion. Its capacitance, based on the *CV* plot, would drop from the maximum value to a mimimum value, such as from point 1 to point 2 in Fig. 4-29b. With the gate voltage held at a steady value, the capacitance would rise slowly until the inversion value C_{min} is reached, i.e., point 3. When capacitance is plotted against time, Fig. 4-29c results.

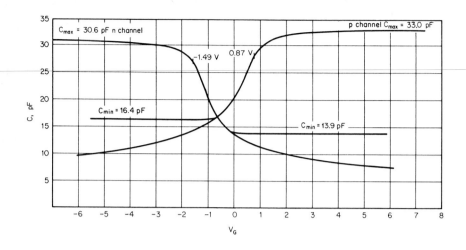

```
ENTER MASK SET NUMBER    AB1234

ENTER LOT NUMBER    RUN#909

ENTER WAFER I.D.    WAF101

ENTER CHANNEL TYPE (N OR P)    N

ENTER GATE MATERIAL: ALUM,N+POLY,P+POLY  (A,N,OR P)    N

ENTER CMAX(PF),CMIN(PF),& AREA(SQ MILS)  30.6,13.9,160

CFB= 26.18 PF
ENTER VOLTAGE FOR CFB -1.49


            CV PLOT ANALYSIS

MASK SET= AB1234
LOT NUMBER= RUN#909
WAFER I.D.= WAF101

                INPUTS:
N CHANNEL        DOT AREA= 160 SQ MILS
CMAX= 30.6 PF    CMIN= 13.9 PF
VFB= -1.49 V     FOR CFB= 26.18 PF
GATE MATERIAL= N+ POLY

                OUTPUTS:
TOX= 1166 ANGSTROMS
NA= 4.79 E+15 CM-3
PHI M-S= -.89 VOLT
VT= .25 VOLT
QB/Q= -2.03 E+11 CM-2
XDMAX= .42 MICRON
NET INTERFACE CHARGE= 11.22 E+10 CM-2
```

Figure 4-28 A typical run using the *CV* analysis program.

```
ENTER MASK SET NUMBER    AB1234

ENTER LOT NUMBER    RUN#909

ENTER WAFER I.D.    WAF101

ENTER CHANNEL TYPE (N OR P)    P

ENTER GATE MATERIAL: ALUM,N+POLY,P+POLY  (A,N,OR P)    P

ENTER CMAX(PF),CMIN(PF),& AREA(SQ MILS)  33.0,16.4,160

CFB= 28.96 PF
ENTER VOLTAGE FOR CFB 0.87


              CV PLOT ANALYSIS

MASK SET= AB1234
LOT NUMBER= RUN#909
WAFER I.D.= WAF101

                INPUTS:
P CHANNEL       DOT AREA= 160 SQ MILS
CMAX= 33 PF     CMIN= 16.4 PF
VFB= .87 V      FOR CFB= 28.96 PF
GATE MATERIAL= P+ POLY

                OUTPUTS:
TOX= 1081 ANGSTROMS
ND= 8.18000001 E+15 CM-3
PHI M-S= .88 VOLT
VT= -1.17 VOLTS
QB/Q= 2.69 E+11 CM-2
XDMAX= .32 MICRON
NET INTERFACE CHARGE= .29 E+10 CM-2
```

Figure 4-28 (cont'd.)

An approximate solution for the capacitance transient response time T can be obtained by the following argument[7]: the generation rate g inside a depletion layer is

$$(4\text{-}20) \quad g = \frac{n_i}{2\tau} \text{ cm}^{-3}/\text{s}$$

where n_i is the intrinsic carrier density, and τ is the effective lifetime. To neutralize a depleted layer of width W, a generation rate per unit area of gW needs to take place in the time T, i.e.,

$$(4\text{-}21) \quad gTW = N_A W$$

Thus

$$(4\text{-}22) \quad T = \frac{N_A}{g} = 2\tau \frac{N_A}{n_i}$$

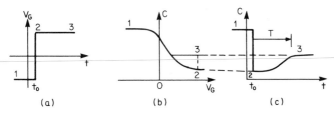

Figure 4-29 A capacitance-vs.-time (*CT*) plot: (*a*) the gate voltage waveform, (*b*) the operating points on the *CV* curve, and (*c*) the resulting *CT* plot.

Since N_A is several orders of magnitude greater than n_i, the capacitance-time plot allows measurement of τ, in microseconds, with T, a time in seconds. Therein lies the utility and popularity of the CT plot. T is typically 10 to 100 s, but with NMOS processes optimized for long lifetimes, figures of 600 to 1000 s (over 10 min) are not uncommon.

Equation (4-22) does not agree with observations that capacitance response time does indeed increase with the magnitude of the applied inversion voltage and with the substrate doping concentration while decreasing with oxide thickness. For that, one should refer to more accurate solutions[8] that describe the whole response plot with

$$(4\text{-}23) \quad \frac{C_o}{C_{\min}} \left[\ln \left(\frac{C_{\min}/C - 1}{C_{\min}/C_o - 1} \right) + \left(\frac{C_{\min}}{C} - \frac{C_{\min}}{C_o} \right) \right] = \frac{-t}{T}$$

where T is still as defined by Eq. (4-22) but is now a normalizing constant rather than the total response time. To determine the lifetime τ from T, one may choose any point on the CT curve and calculate

$$(4\text{-}24) \quad T = \frac{1}{\text{slope of } CT \text{ plot}} \frac{C^2}{C_o} \left(1 - \frac{C}{C_{\min}} \right)$$

A graphical technique is also available.[9]

In performing a CT measurement, even more so than CV plots, one must have MOS capacitors that are free of moisture on the top surface, otherwise the area surrounding the capacitor will also be depleted and contribute generation current. This would give a falsely lower measured lifetime. The CT measurement is also particularly sensitive to light.

Gate-Controlled Diode

An alternative way to measure minority carrier lifetime is by means of the gate-controlled diode structure. The technique is particularly useful in separating the surface and bulk components of thermal carrier generation that takes place in a depleted surface region. The gate-controlled diode structure consists of a diode with a gate electrode surrounding its periphery. A cross section of the structure is shown in Fig. 4-30. The technique requires a fixed voltage V_R to be applied to reverse-bias the diode. The reverse diode current is then plotted as a function of a different voltage that is applied *to the gate*. If the gate voltage causes the underlying silicon to be in accumulation, the reverse current would stay constant and would equal the normal reverse current of the diode I_R. When the gate voltage is increased and begins to deplete the silicon surface, the depletion layer under the gate couples to the junction depletion layer and the reverse-bias current increases significantly. Further increase in the gate voltage will increase the surface depletion-layer width and with it the reverse current. The reason for this is that the reverse-biased diode continually drains away any generated carriers and the surface of the depletion layer stays fully depleted. Bulk traps in the depletion layer's entire volume contribute to the bulk component of the generated current, while interface states contribute the surface component. When the gate voltage is increased even further until the surface potential under the gate is at $V_R + 2\phi_f$, the surface will invert. The surface of the depletion layer is no longer depleted but has a high carrier concentration. This quenches the surface generation, causing an abrupt drop in current. That drop is the surface component I_S, and the remainder of the excess current is the bulk component I_B.

To construct a gate-controlled diode, the gate area needs to be several times larger than the diode area to ensure that the surface component will be large and that the drop in current will be quite distinct. The structure is also often made of concentric circles to eliminate fringing effects.

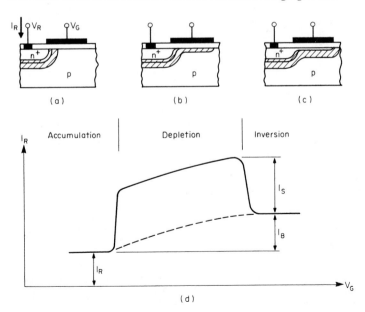

Figure 4-30 A gate-controlled diode.

REFERENCES

1. K. H. Zaininger and F. P. Heiman, "The *C-V* technique as an analytical tool," *Solid-State Technology,* part I, pp. 49–56, May 1970; part II, pp. 46–55, June 1970.

2. S. M. Sze, *Physics of Semiconductor Devices,* 2d ed., Chap. 8, Wiley, New York, 1981.

3. A. Goetzberger, "Ideal MOS curves for silicon," *Bell Sys. Tech. J.,* vol. 45, p. 1097, 1966.

4. B. E. Deal, E. H. Snow, and C. A. Mead, "Barrier energies in metal-silicon dioxide-silicon structures," *J. Phys. Chem. Solids,* vol. 27, p. 1873, 1966.

5. E. H. Snow, A. S. Grove, and D. J. Fitzgerald, "Effects of ionizing radiation on oxidized silicon surfaces and planar devices," *Proc. IEEE,* vol. 55, pp. 1168–1185, July 1967.

6. P. Burggraaf, "*C-V* plotting, *C-T* measuring and dopant profiling: Applications and equipment," *Semiconductor Int.,* pp. 29–42, October 1980.

7. J. Grosvalet, C. Jund, et al., "Experimental study of semiconductor surface conductivity," *Surface Science,* vol. 5, p. 49, 1966.

8. F. P. Heiman, "On determination of minority carrier lifetime from the transient response of a MOS capacitor," *IEEE Trans. Electr. Dev.,* vol. ED-14, p. 781, 1967.

9. R. F. Pierret, "Rapid interpretation of the MOS-C C-T transient," *IEEE Trans. Electr. Dev.,* vol. ED-25, pp. 1157–1159, September 1978.

10. B. Deal, "Standardized terminology for oxide charges associated with thermally oxidized silicon," *IEEE Trans. Electr. Dev.,* vol. ED-27, no. 3, pp. 606–608, March 1980.

11. B. Deal, "The current understanding of charges in the thermally oxidized silicon structure," *J. Electrochem. Soc.,* vol. 121, no. 6, pp. 198C–205C, June 1974.

PROBLEMS

1. If a MOS capacitor is contaminated with mobile ions in the oxide, prove that the shift in the *CV* plot is in the opposite direction of the applied bias, regardless of the sign of the ionic charge.

2. (a) The thicker the gate oxide, the (larger/smaller) the threshold voltage. The higher the background concentration, the (larger/smaller) the threshold voltage.
   (b) Q_f (increases/decreases) the threshold voltage of an n-channel device and (increases/decreases) the magnitude of the threshold of a p-channel device.

3. In the section on *CT* plots, it is stated that if a voltage greater than the threshold is suddenly applied to the MOS capacitor, one would wait anywhere from a few to hundreds of seconds before inversion is reached. By analogy, when a MOS IC

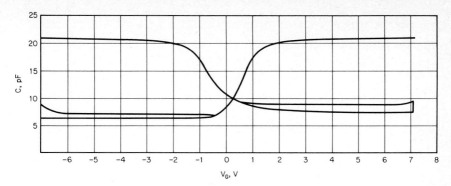

Figure 4-31 Problem 4.

is first powered up, one should wait a similar time for the inversion layers under the transistor to form. True or false. Why?

4. Run the *CV* analysis program for the two *CV* plots in Fig. 4-31. The *n*-channel device uses n^+ poly, while the *p*-channel device uses p^+ poly. Invent your own identifier for the mask set, lot number, or wafer identification.

Appendix
CV Analysis Program

```
100  REM      CV ANALYSIS PROGRAM
110  REM    THIS PROGRAM CALCULATES OXIDE THICKNESS,
120  REM    DOPING CONCENTRATION, WORK FUNCTION, FLAT
130  REM    BAND CAPACITANCE, DEPLETION CHARGE,
140  REM    DEPLETION LAYER WIDTH, THRESHOLD, AND
150  REM    INTERFACE CHARGE DENSITY USING C-V PLOT.
160  REM
170  PRINT
180  PRINT
190  INPUT "ENTER MASK SET NUMBER    ";MASK$
200  PRINT
210  INPUT "ENTER LOT NUMBER    ";LOT$
220  PRINT
230  INPUT "ENTER WAFER I.D.    ";WAFER$
240  PRINT
250  INPUT "ENTER CHANNEL TYPE (N OR P)    ";ISUB$
260  PRINT
270  SUBS = 1
280  IF ISUB$ = "P" THEN SUBS = - 1
290  INPUT "ENTER GATE MATERIAL: ALUM,N+POLY,P+POLY   (A,N,OR P)
";MET$
300  PRINT
310  INPUT "ENTER CMAX(PF),CMIN(PF),& AREA(SQ MILS)   ";CO,CMIN,AD
320  DOX = 223.0 * AD / CO
330  NAD = 5.0E15
340  NX = NAD
350  TEMP = (CMIN / CO) / (1 - (CMIN / CO))
360  NAD = (7.4E20 / (DOX ^ 2)) * TEMP ^ 2 *  LOG (NX / 1.45E10)
370  TEMP =  ABS ((NAD - NX) / NAD)
380  IF TEMP > 0.01 THEN  GOTO 340
390  PHF = 0.0259 *  LOG (NAD / 1.45E10) * SUBS
400  IF MET$ = "A" THEN M = - 0.595
410  IF MET$ = "P" THEN M = 0.542
420  IF MET$ = "N" THEN M = - 0.555
430  PIMS = M - PHF
440  CSFB = 1.639E - 8 *  SQR (NAD) * AD
450  CFB = CSFB * CO / (CSFB + CO)
460  IF SUBS < 0 THEN QBQ = 3.6E3 * ((2 *  ABS (PHF) * NAD) ^
0.5)
470  IF SUBS > 0 THEN QBQ =  - 3.6E3 * ((2 * PHF * NAD) ^ 0.5)
480  XDMAX =  ABS (QBQ) * 1E4 / NAD
490  VTVFB = 2 * PHF + SUBS * 1.032E - 12 *  ABS (QBQ) * AD / CO
500  CFB = 0.01 *  INT (CFB * 100)
510  PRINT
520  PRINT "CFB= ";CFB;" PF"
530  INPUT "ENTER VOLTAGE FOR CFB ";VFB1
540  VVT = VTVFB + VFB1
550  QSSQ1 = 9.69E11 * CO * (PIMS - VFB1) / AD
560  PRINT
570  PRINT
580  PRINT
590  PRINT "             CV PLOT ANALYSIS"
600  PRINT
610  PRINT "MASK SET= ";MASK$
620  PRINT "LOT NUMBER= ";LOT$
630  PRINT "WAFER I.D.= ";WAFER$
```

```
640   PRINT
650   PRINT "                    INPUTS:"
660 AD = 0.01 *  INT (AD * 100)
670   PRINT ISUB$;" CHANNEL","DOT AREA= ";AD;" SQ MILS"
675   PRINT "CMAX= ";CO;" PF","CMIN= ";CMIN;" PF"
676 CFB = 0.01 *  INT (CFB * 100)
677   PRINT "VFB= ";VFB1;" V","FOR CFB= ";CFB;" PF"
680   IF MET$ = "A" THEN  PRINT "GATE MATERIAL= ALUMINUM"
690   IF MET$ = "N" THEN  PRINT "GATE MATERIAL= N+ POLY"
700   IF MET$ = "P" THEN  PRINT "GATE MATERIAL= P+ POLY"
730   PRINT
740   PRINT "                    OUTPUTS:"
750 DOX =  INT (DOX)
760   PRINT "TOX= ";DOX;" ANGSTROMS"
770 NAD = 0.01 *  INT (NAD / 1.E15 * 100)
780   IF ISUB$ = "N" THEN  PRINT "NA= ";NAD;" E+15 CM-3"
790   IF ISUB$ = "P" THEN  PRINT "ND= ";NAD;" E+15 CM-3"
800 PIMS = 0.01 *  INT (PIMS * 100)
810   PRINT "PHI M-S= ";PIMS;" VOLT"
820 VVT = 0.01 *  INT (VVT * 100)
830   PRINT "VT= ";VVT;" VOLTS"
840 QBQ = 0.01 *  INT (QBQ / 1.E11 * 100)
850 XDMAX = 0.01 *  INT (XDMAX * 100)
860   PRINT "QB/Q= ";QBQ;" E+11 CM-2"
870   PRINT "XDMAX= ";XDMAX;" MICRON"
880 QSSQ1 = 0.01 *  INT (QSSQ1 / 1.E10 * 100)
890   PRINT "NET INTERFACE CHARGE= ";QSSQ1;" E+10 CM-2"
900   PRINT
910   END
```

5

MOS Device Physics

5-1 TRIODE REGION

This chapter will cover in greater detail the various operating regions of the MOS device characteristics. First to be discussed is the triode region.

The transfer characteristics (I_D versus V_D) of a MOS transistor operated in the triode (sometimes called linear) region will be derived in this section. The n-channel device will again be used as the basis for analysis.

In the triode region, the drain voltage is small relative to the gate voltage, and as depicted in Fig. 5-1, there is a continuous inversion channel extending from the source to the drain. The channel has a constant width W, a constant length L, and a thickness x_c that changes somewhat from source to drain. The channel is redrawn on the right half of the figure, defining the current to be flowing in the y direction. By taking a thin slice of the channel perpendicular to the current flow, one can write down the differential voltage drop across it in the y direction:

$$(5\text{-}1) \quad dV = I_D \, dR$$

But

$$(5\text{-}2) \quad dR = \frac{\rho \, dy}{x_c(y) W}$$

where $x_c(y)$ changes as a function of y. In addition

$$(5\text{-}3) \quad \rho = (q\mu n)^{-1}$$

Thus

$$(5\text{-}4) \quad dV = \frac{I_D \, dy}{q\mu n x_c(y) W}$$

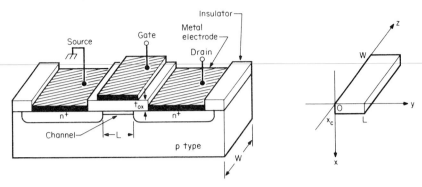

Figure 5-1 The reference directions for the analysis of a MOS transistor. (*After Ref. 3-2.*)

But the inversion charge is

(5-5) $\quad qx_c(y)n = Q_n(y)$

Hence

(5-6) $\quad dV = \dfrac{I_D \, dy}{Q_n(y)\mu W}$

and

(5-7) $\quad I_D \, dy = Q_n(y)\mu W \, dV$

Since the drain voltage is dropped across the channel from drain to source, the voltage must vary along the channel. For the same gate voltage, the inversion-layer charge would vary with the voltage across the oxide; thus

(5-8) $\quad Q_n(y) = [V_G - V_T - V(y)]C_o$

Equation (5-7) then becomes

(5-9) $\quad I_D \, dy = [V_G - V_T - V(y)]C_o\mu W \, dV$

Integrating from source to drain requires integrating y from 0 to L and V from 0 to V_D:

(5-10) $\quad I_D \displaystyle\int_0^L dy = C_o\mu W \int_0^{V_D} [(V_G - V_T) - V(y)] \, dV$

(5-11) $\quad I_D = C_o\mu \dfrac{W}{L}[(V_G - V_T)V_D - \tfrac{1}{2}V_D^2]$

Thus

(5-12) $\quad \boxed{I_D = \beta[(V_G - V_T)V_D - \tfrac{1}{2}V_D^2]}$

where

(5-13) $\beta = \beta_o \dfrac{W}{L}$

and

(5-14) $\beta_o = C_o \mu = \dfrac{\epsilon_{ox}\epsilon_o}{t_{ox}} \mu$

The device cross section when in the triode region is shown in Fig. 5-2*a*. Note the depletion layer that surrounds the source junction (even at $V_A = 0$), inversion channel, and drain junction (where $V_A = V_D$).

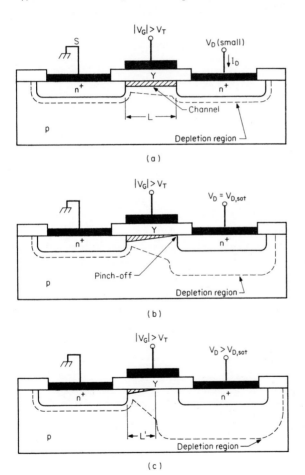

Figure 5-2 Device cross section in (*a*) triode region, (*b*) onset of saturation, and (*c*) beyond saturation. (*After Ref. 3-2.*)

5-2 SATURATION REGION

Starting from zero current, if the drain voltage is increased, the drain current increases, but the rate of rise falls off until eventually the drain current saturates to a constant value. This behavior is illustrated in Fig. 5-3. The area where the drain current is no longer affected by the drain voltage, and where the device acts more like a current source, is called the *saturation region*. The dividing line between triode and saturation regions is where the voltage drop across the gate oxide falls below V_T, causing the channel to pinch off. Note that as the drain voltage increases, the voltage across the oxide is decreasing. Because the drain voltage is dropped from drain to source, this pinch off always occurs first at the drain junction. The pinch-off condition is then described by

(5-15) $V_G - V_{D,\text{sat}} = V_T$

or

(5-16) $V_{D,\text{sat}} = V_G - V_T$

The current is then

(5-17) $\boxed{I_D = \tfrac{1}{2}\beta(V_G - V_T)^2}$

The device cross section when in saturation is shown in Fig. 5-2b. The large drain voltage increases the drain junction reverse bias and widens the depletion layer. Along the channel, the voltage drops from V_D to 0; the depletion-layer width under the gate correspondingly drops from being equal to the drain junction depletion width, down to the equilibrium $\phi_s = 2\phi_f$ value. Since there is current flow, the device is no longer in equilibrium, and the depletion layer under the gate can be greater than the "maximum" depletion width defined for the equilibrium condition of $\phi_s = 2\phi_f$.

When the drain voltage is beyond onset of saturation, the pinch-off point starts to move toward the source (Fig. 5-2c). This movement is referred to

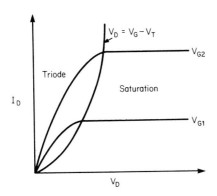

Figure 5-3 The triode and saturation regions of operation, and the dividing line between the two.

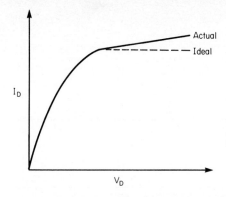

Figure 5-4 The nonzero slope caused by the channel shortening effect.

as *channel length modulation.* The pinch-off point, Y in the figure, still has a voltage $V_{D,sat}$ as described by Eq. (5-16), but the portion of the drain voltage beyond $V_{D,sat}$ is now dropped across the depletion layer from point Y to the drain. Since the same voltage $V_{D,sat}$ is now dropped across a shorter channel length L', the resultant drain current will increase slightly. The higher the drain voltage, the closer the pinch-off point will move toward the source, and the higher the current. This places a slope on the characteristic curve as shown in Fig. 5-4. The nonzero slope represents a finite output impedance when the device is used in analog circuits and an extra current drain when the device is used in digital circuits. The change in drain current is most severe when the change in channel length $L - L'$ is a large fraction of the initial channel length L.

The saturated current with channel length modulation taken into account can be approximated by

$$(5\text{-}18) \quad I_D = \frac{I_{D,sat}}{L'/L}$$

where $L - L'$ can be calculated by assuming the voltage of $(V_D - V_{D,sat})$ is dropped across a *pn* junction

$$(5\text{-}19) \quad L - L' = \left[\frac{2\epsilon_s\epsilon_o(V_D - V_{D,sat})}{qN_A} \right]^{1/2}$$

The fact that the channel terminates at the pinch-off point Y does not shut off the current. Rather, any carriers that travel from the source via the channel will simply be injected into the drain depletion region, identical to the conduction of current through the depletion regions of a *pn* junction.

When the MOS transistor is configured with its gate tied to the drain, as in Fig. 5-5, then the device is constrained to operate in the saturation region. Its two-terminal $I_D - V_D$ characteristic is a square-law type function that lies always to the right of the $V_D = V_G - V_T$ locus that separates the triode and saturation regions. Alternatively, one can argue that since V_G

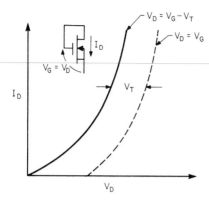

Figure 5-5 $I_D - V_D$ characteristics of a transistor with its gate
tied to its drain.

$= V_D$, the voltage across the oxide is always zero and the channel region
near the drain has to be pinched off. This special situation holds true only
for an enhancement-mode device. For a depletion-mode device tying the gate
to the drain yields an $I_D - V_D$ curve that stays in the triode region and
passes through the origin, but in a manner that holds no particular interest.

5-3 AVALANCHE REGION

As the drain voltage is increased further from the saturation region, a point
is reached where the drain current increases abruptly. The device enters a
breakdown mode, and this area of operation is called the *avalanche region*.
Unlike bipolar transistors, if the drain voltage is stable and is not actually
right at the breakdown voltage, the MOS transistor does not go into thermal
runaway.

The drain junction is in reverse bias under normal operation; but its
breakdown voltage is lower than that predicted by a simple *pn* step junction.
As explained in Chap. 2, this is due largely to the curvature that results
from diffusing junctions in integrated circuit processing. However, it is
found that the drain breakdown voltage is even lower than that attributed
to the junction curvature effect. For the explanation, one should refer to Fig.
5-6. The presence of the gate electrode results in high field regions in the
drain junction, lowering its breakdown voltage. This was found to be true
whether the gate was left floating or was grounded. When a gate voltage is
applied, it is always in the direction to lower the gate-to-drain voltage difference across the oxide. In effect, low gate voltage curves will exhibit lower
drain breakdown, resulting in characteristics similar to Fig. 5-7. Such characteristics are typical of metal gate devices which require a large overlap
between the gate and drain. For silicon gate devices where the drain self-
aligns to the gate (see Sec. 7-4), the drain breakdown characteristics are
more of the type shown in Fig. 5-8. They are similar to bipolar transistor

breakdown characteristics and are the result of carrier multiplication, i.e., larger current induces earlier breakdown.

There is a special case of high drain current that is called *punch-through*. This happens when the drain depletion layer actually touches the source depletion layer and space-charge–limited current flows from source to drain. This condition is most likely to be encountered in devices with either very

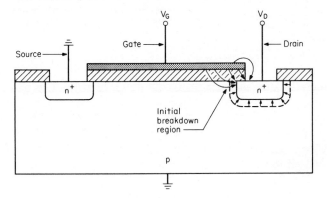

Figure 5-6 Enhanced field regions at the drain.

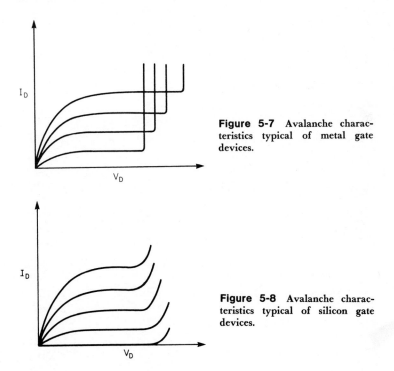

Figure 5-7 Avalanche characteristics typical of metal gate devices.

Figure 5-8 Avalanche characteristics typical of silicon gate devices.

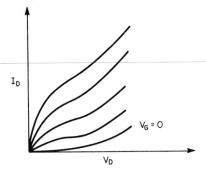

I_D

$V_G = 0$

V_D

Figure 5-9   **Punch-through char-acteristics typical of short-channel devices.**

short channel lengths, high resistivity substrates, or both. The resulting characteristics of Fig. 5-9 are quite typical. The most striking feature of the figure is that even with $V_G = 0$, there is substantial current flow at low values of drain voltage. When a gate voltage that is greater than the threshold is applied, a surface channel is formed as usual, and the normal drift current is added to the space-charge–limited current flowing through the depleted region. The gate, in effect, still modulates the drain current, as shown by Fig. 5-9. However, when reverse voltage is applied to the gate, shutting the surface channel off still leaves the underlying depletion-layer current unaffected. The device then cannot be completely shut off. Only with the application of a reverse source-to-substrate potential can the device be shut off for a usable range of drain voltages.[1]

5-4   SUBTHRESHOLD REGION

When the gate voltage of a MOS transistor is reduced to V_T, the current does not abruptly drop to zero. Figure 5-10 shows that right at V_T, or even below V_T, there is still current flow. If several orders of magnitude of the drain current are plotted with respect to V_G for a given V_D, then Fig. 5-11 is a typical result. I_D drops off in an exponential manner, with a slope that causes a tenfold change in current for every 100 to 200 mV. The curved section of the graph is where the device enters the triode or saturation region. For other values of V_D, the plot is observed to be vertically translated. Because of the steep slope, this results in a lateral translation of only fractions of a volt.

Analysis[2] has shown that the subthreshold behavior can be predicted by

$$(5\text{-}20) \quad I_D = \frac{W}{L} \mu_n C_o \frac{1}{m} \left(n \frac{kT}{q} \right)^2 \exp \left[\frac{q}{nkT} \left(V_G - V_T - n \frac{kT}{q} \right) \right]$$

$$\times \left[1 - \exp \left(\frac{-mq}{nkT} V_D \right) \right] \simeq I_o \exp \left[q \frac{(V_G - V_T)}{nkT} \right] \qquad V_G \leq V_T$$

where $m = (C_o + C_s)/C_o$ and $n = (C_o + C_s + C_{fs})/C_o$. C_o is the oxide capacitance, and C_{fs} is the equivalent capacitance when surface states charge and discharge carriers with changes in surface potential. Typical values for m and n range from 1.5 to 3.0. Equation (5-20) does predict the exponential dependence of the drain current on V_G and the weak dependence on V_D.

An important consequence of Eq. (5-20) is that it places a lower limit on reductions in threshold voltages for low-voltage (1- to 2-V) operation. In order to have a distinction between on-off states of a transistor, the current needs to change by three to four orders of magnitude. Equation (5-20) predicts that the threshold voltage should be set high enough to allow at least

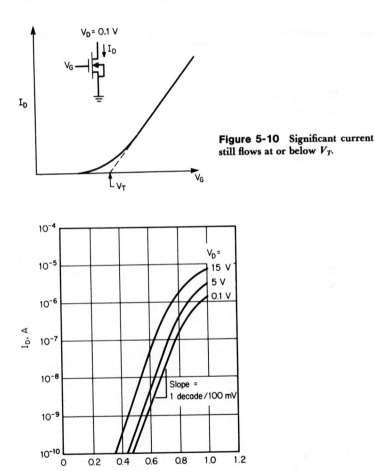

Figure 5-10 Significant current still flows at or below V_T.

Figure 5-11 Plot of drain current over several orders of magnitude as a function of gate voltage for several given values of V_D.

a 10 kT/q change in gate voltage.[3] At room temperature, V_T is then 250 mV and the power supply voltage should be about three times that or 750 mV. Another consequence is that when a MOS transistor is used to isolate the charge stored on a capacitor, its gate voltage should be forced far below V_T to ensure low leakage.

5-5 SECOND-ORDER EFFECTS

Nonconstant Surface Mobility

The surface mobility of a MOS transistor depends on the surface roughness and on the number of interface traps present which can trap and retard carriers. For typical MOS processes, surface mobility values range from 150 to 250 cm^2/V·s for holes and 550 to 950 cm^2/V·s for electrons. The ratio of electron-to-hole mobility ranges from 2 to 4. This fact is utilized in CMOS circuit design, where in order to have equal n-channel and p-channel transconductance in an inverter, p-channel devices have to be two to four times wider than n-channel devices.

The mobility values just quoted are measured at low gate field, i.e., with the gate voltage only 1 to 2 V above threshold. At higher gate voltages, it is found that mobility is often reduced. With larger gate fields, the current carriers encounter more collisions with the oxide-semiconductor interface as it drifts along the channel. This results in lower carrier mobility and manifests itself in several ways. It shows up in the transfer characteristics where instead of the saturation drain current increasing in a square-law manner with gate voltage (see Fig. 1-6), the constant V_G curves are much closer together. In the triode region, the curves at high gate voltage start to bunch up or crimp together instead of being clearly spaced apart.[1] This phenomenon also shows up in the triode conductance curve (Fig. 5-12) where the drain current is plotted as a function of gate voltage. The small value of drain voltage, usually 0.1 V, forces the device to be in the triode region. The name conductance curve comes from the fact that when the current is divided by the fixed drain voltage, a plot of conductance is obtained that no

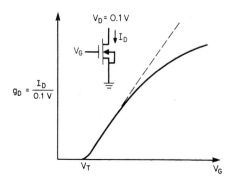

Figure 5-12 Conductance curve showing reduction in mobility at high gate voltage.

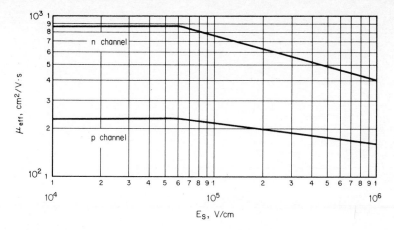

Figure 5-13 Mobility reduction as a result of high field.

longer is influenced by the actual value of V_D used. Figure 5-12 shows that the conductance deviates from the ideal linear relationship at large gate voltage, the reduction in conductance being caused by a reduction in mobility. Empirical data show that the mobility is constant at low field and starts falling off at around 0.5 to 1×10^5 V/cm, such as in Fig. 5-13. This suggests a model of the form

$$(5\text{-}21) \quad \frac{1}{\mu} = \frac{1}{\mu_o} + \text{FRC}\,\frac{V_G - V_T}{t_{ox}}$$

where μ_o = low-field mobility

FRC = field reduction coefficient, units of s/cm

$(V_G - V_T)/t_{ox}$ = vertical field across oxide

To measure μ_o and FRC, one may use a conductance curve such as Fig. 5-12 and calculate $1/\mu$ for several points along the curve and plotted against $V_G - V_T$. $1/\mu$ is calculated from

$$(5\text{-}22) \quad \frac{1}{\mu} = C_o \frac{W}{L} \frac{V_G - V_T}{g_D}$$

When plotted, Fig. 5-14 shows a typical result. A straight line can be fitted to the curve. It can be seen from Eq. (5-21) that the intercept of the curve would yield $1/\mu_o$, and the slope of the line yields the field reduction coefficient multiplied by t_{ox}.

Back-Gate Bias

Although most devices are operated with their source grounded to the substrate, there are numerous situations where there is a voltage difference

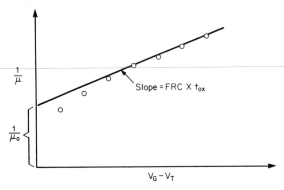

Figure 5-14 Graph for calculating low-field mobility μ_o and field reduction coefficient.

between the two. Figure 5-15 shows cases where this condition can occur. One is where the substrate is at ground and the source voltage can rise above ground. This is true of MOS transistors that are operated as load devices in an inverter. Their sources are the outputs of the inverter and will certainly swing above ground. The other situation, actually electrically equivalent, is when all sources are grounded and an external voltage source is applied to the substrate. In both cases, the back-gate bias V_{BG} reverse-biases the source with respect to the substrate and causes an increase in the threshold voltage of the device. In the case of p-channel devices, back-gate bias also reverse-biases the source-substrate junction and increases the magnitude of its threshold voltage. Threshold voltage is increased because the back-gate bias widens the depletion layer under the gate and adds to the number of charges the gate voltage has to balance before the inversion condition is achieved.

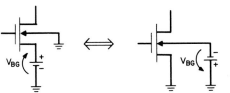

Figure 5-15 Back-gate bias between source and substrate.

The depletion-layer width now corresponds to a surface potential of $2\phi_f + V_{BG}$ instead of just $2\phi_f$. The threshold voltage is thus

(5-23) $$V_T = V_{FB} + 2\phi_f + \frac{\sqrt{2\epsilon_s\epsilon_o qN_A(2\phi_f + V_{BG})}}{C_o}$$

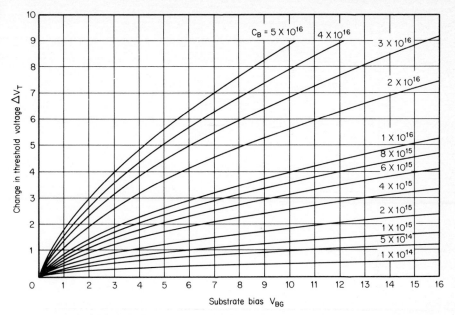

Figure 5-16 Increase in threshold voltage as a function of back-gate bias t_{ox} = 1000 Å. For other values of t_{ox}, scale ΔV_T by $t_{ox}/1000$ Å. p channel: V_{BG} is + and ΔV_T is −. n channel: V_{BG} is − and ΔV_T is +.

In quantifying the effect of back-gate bias, it is useful to retain the usual V_T value and calculate ΔV_T as the increase in threshold due to the effect. ΔV_T is simply the difference between Eq. (5-23) and the usual V_T expression given by Eq. (3-13) for $V_{BG} = 0$:

$$\text{(5-24)} \quad \Delta V_T = \frac{\sqrt{2\epsilon_s\epsilon_o q N_A}}{C_o}\left(\sqrt{2\phi_f + V_{BG}} - \sqrt{2\phi_f}\right)$$

A plot of ΔV_T as a function of V_{BG} is shown in Fig. 5-16 for a given oxide thickness and for various background doping concentrations. It is evident from Eq. (5-24) that the back bias effect is larger for thicker gate oxides. Also confirmed by Fig. 5-16, the effect is worse for larger doping concentrations.

Equation (5-24) has also been used for calculating the background doping concentration profile. If the conductance plot is made for various back-gate biases, as in Fig. 5-21, the curve will translate laterally by an amount equivalent to ΔV_T. If ΔV_T is plotted as a function of $\sqrt{2\phi_f + V_{BG}}$, there results a straight line whose slope can be used to calculate N_A if the oxide thickness is known. To be more precise, $2\phi_f$ will change with N_A, but the relationship is a logarithmic one, and the magnitude of $2\phi_f$ is small when compared with typical V_{BG}; thus little error will result by treating it as a

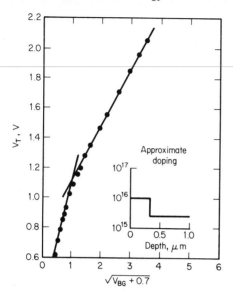

Figure 5-17 Different doping concentrations show up as different slopes in a V_T vs. $\sqrt{2\phi_f + V_{BG}}$ plot.

constant. If the substrate doping concentration has been raised by ion implantation and the doping profile can be approximated by two rectangles (see insert in Fig. 5-17), then with this technique, a break in the slope of the ΔV_T versus $\sqrt{2\phi_f + V_{BG}}$ can be observed.

Varying Bulk Charge

The derivation of the device equation in Sec. 5-1 assumes a constant depletion-layer width underneath the channel that is everywhere equal to its value at the source end. This is close to the true situation for small V_{DS}. Such a model, though simple, yields compact device equations and allows definition of a single threshold voltage for a transistor. However, as V_{DS} increases, and particularly when V_D approaches the saturation value $V_{D,\text{sat}}$, then the depletion layer under the channel is certainly not constant (see Fig. 5-2b). The device equations can be re-derived to take the varying bulk charge into account.

In the following analysis as well as in the simple model case, the *gradual-channel* approximation is invoked. It assumes that the field in the direction perpendicular to the silicon surface is much greater than the field in the direction of current flow, and allows the use of one-dimensional MOS analysis.

The inversion-layer charge $Q_n(y)$ given by Eq. (5-8) is simply rewritten with all the components of V_T put back in, i.e., Eqs. (3-13) and (3-22). In particular, the bulk charge Q_B now changes with y, the distance along the channel. Instead of having a constant surface potential $2\phi_f$, Q_B now has a potential $2\phi_f + V(y)$ where $V(y)$ is the voltage drop along the channel. A separation of variables is made as before, with $V(y)$ integrated from 0 to V_D and y from 0 to L. The net result is

(5-25)
$$I_D = \beta \left\{ \left(V_G - V_{FB} - 2\phi_f - \frac{V_D}{2} \right) V_D \right.$$
$$\left. - \frac{2}{3} \frac{\sqrt{2\epsilon_s\epsilon_o qN_A}}{C_o} \times [(V_D + 2\phi_f)^{3/2} - (2\phi_f)^{3/2}] \right\}$$

One consequence of Eq. (5-25) is that there is no longer a single quantity defined as V_T. This makes the equation more difficult to use in calculations but is found to agree better with actual devices. Figure 5-18 compares the device characteristics as predicted by the simple model to those of the varying bulk charge model. The simple model overestimates the current by 20 to 50 percent and predicts a larger $V_{d,sat}$ (shown as tick marks on the curves). As expected, the two curves converge at low V_D. This is observed to occur for I_D less than roughly 20 percent of its maximum value.

The saturation voltage is calculated by letting the channel pinch off, i.e., $Q_n(L) = 0$. This leads to

(5-26) $\quad V_{D,sat} = V_G - V_{FB} - 2\phi_f$
$$- \frac{\epsilon_s\epsilon_o qN_A}{C_o^2} \left[\sqrt{1 + \frac{2C_o^2}{\epsilon_s\epsilon_o qN_A}(V_G - V_{FB})} - 1 \right]$$

5-6 WAYS OF MEASURING THRESHOLD VOLTAGE

The threshold voltage is one of the most important parameters of the MOS transistor and can be described simply as the value of the gate voltage when current starts to flow between the source and drain. Yet there are many ways of measuring threshold that are in accepted usage that will yield slightly different values. The reason is that the device itself does not exhibit

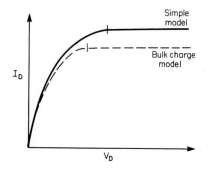

Figure 5-18 Comparison between simple model and varying bulk charge model. The tick marks are the drain saturation voltages.

a well-defined cutoff, as Sec. 5-4 pointed out. The different ways of measurement all give the same threshold to within 100 to 200 mV. But often when more precision is required, the V_T value quoted must be identified by the method by which it was measured. With all the methods to be described below, with the exception of the last one, V_T is measured with the device in saturation.

1-μA Method

V_T is arbitrarily defined as the gate voltage when 1 μA of drain current flows. This method came into popular use in the early days of MOS work since it is straightforward. All it involves is to note down on a curve tracer the gate voltage at which the flat saturation drain current reaches 1 μA. Its major drawback is that there is no way to take into account the size, i.e., the W/L ratio, of the device. Two different-sized devices side by side on the same wafer will yield different thresholds if this method is used. For this reason, the method is no longer recommended.

$\sqrt{I_D}$ versus V_G Method

If the square root of the drain current is plotted as a function of gate voltage for a saturated device, a straight line will result as in Fig. 5-19, whose intercept with the X axis is V_T. One way to ensure that the device is in saturation is to tie the gate to the drain. The validity of this method is seen from taking the square root of both sides of the saturation current equation:

$$(5\text{-}27) \qquad \sqrt{I_D} = \sqrt{\frac{\beta}{2}} (V_G - V_T)$$

$I_D = 0$ when $V_G = V_T$. The slope of the curve is

$$(5\text{-}28) \qquad \frac{d\sqrt{I_D}}{dV_G} = \sqrt{\frac{\beta}{2}}$$

This method will work for both enhancement- and depletion-mode devices, although recall that forcing a device into saturation by tying its gate to the drain works only with enhancement devices.

Actual devices will not exhibit the sharp intercept predicted by Eq. (5-27) because of weak inversion current; thus graphic extrapolation from the *linear* portion of the curve is needed. Different size devices adjacent to each other will produce lines that extrapolate to the same point when this method is used.

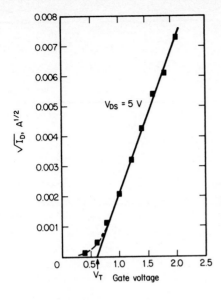

Figure 5-19 Plotting the square root of I_D vs. V_G of a saturated device yields a straight line whose intercept is V_T.

10–40 Method

This method is a variation of the previous method, but it eliminates the graphic extrapolation and replaces it with a simple calculation. A gate voltage is applied and adjusted until 10 μA of saturation drain current flows, measured at, say, $V_D = 5$ V. This gate voltage is called V_{10}. The gate voltage is then increased until 40 μA flows. This voltage is called V_{40}. The threshold is easily calculated from

(5-29) $$V_T = 2V_{10} - V_{40}$$

The derivation of Eq. (5-29) is left as an exercise for the reader as part of the problem set.

This method is particularly suited to manual measurements with a curve tracer because most curve tracers have vernier knobs that permit rapid adjustment and precise readout of gate voltages.

It should be pointed out that there is nothing unique about the choice of 40 μA and 10 μA. The two values need only be related by a ratio of 4. The first value 10 μA is chosen to ensure that the device is operating in the linear portion of Fig. 5-19.

Alternative 10–40 Method

This method uses the same concept and formula as the 10–40 method, except that the device is forced into saturation by tying the gate to the drain. The heavy line in Fig. 5-20 shows the two terminal characteristics that will

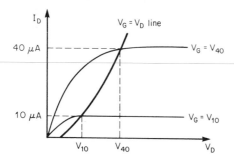

Figure 5-20 V_{10} and V_{40} are the gate voltages for which 10 and 40 μA, respectively, of drain current flows.

appear on a curve tracer and the definition of V_{10} and V_{40}. This method is easier to implement on automatic testers, since they can be programmed to force a current rapidly through the device and measure the corresponding voltage drop. In the original 10–40 method, three terminals are involved. A gate voltage is applied and the drain current measured. If I_D is not within the desired window, the gate voltage is modified and another I_D reading is taken. A search routine has to be implemented to converge to the right gate voltage. This consumes much more machine test time than a two-terminal measurement. The transfer characteristics, shown as light solid lines, for $V_G = V_{40}$ and $V_G = V_{10}$ are superimposed on Fig. 5-20 for reference.

This method is listed separately because it can yield a different threshold from the previous one. If there is no channel length modulation in the saturation region and the drain current has zero slope, such as that drawn for Fig. 5-20, then one gate voltage will yield the same current for either $V_D = V_G$ or $V_D = 5$ V. Then both the 10–40 method and the alternative 10–40 method will yield the same V_T. But if the saturation characteristics have a slope, then to reach, say, 40 μA would require a different $V_G = V_{40}$, depending on what drain voltage it is measured at. This results in threshold voltages that differ by a few hundred millivolts.

Conductance (Triode) Method

This method uses the conductance measurement where drain current is plotted as a function of a varying gate voltage, with V_D fixed at a low value, typically 100 mV. The setup and resulting plot is shown in Fig. 5-21. The small drain voltage forces the device into the triode region and also permits dropping the V_D^2 term in the triode equation. Thus

$$(5\text{-}30) \quad g_D = \frac{I_D}{V_D} = \beta(V_G - V_T)$$

The slope is

$$(5\text{-}31) \quad \frac{dI_D/V_D}{dV_G} = \beta$$

The graph will show a straight line region whose intercept with the X axis is V_T. By applying a back-gate bias, the threshold will increase in the manner described in "Back-Gate Bias," in Sec. 5-5.

This method is by far the most common one used for devices with short channel length because the small drain voltage minimizes the channel shortening effect. But this method is also difficult to automate. The problem lies in finding the proper linear region to extrapolate from. Near V_T, the device exhibits nonlinearities due to conduction in weak inversion. At large gate voltages, the curve concaves downward due to the field reduction of mobility. Finding the inflection region where the curve is linear presents a challenge for simple testers, particularly if even the range of that linear region is not known. Solutions often take the form of customizing the test program after typical conductance plots are taken manually. One example is simply to take readings over a 2-V range after the current exceeds a point known to be in the linear region.

5-7 MOS DEVICE APPLICATIONS

Depletion-Mode Device

A depletion-mode device is one which has a built-in channel, even with zero-applied gate voltage. Current can then flow from the source to the drain, as Fig. 5-22a illustrates. With increasing gate voltage (positive for n-channel devices), current increases. With the opposite polarity gate voltage, the drain current is reduced until the channel is pinched off at $V_G = V_T$. Coming from the other direction, one could say that no current flows until V_T is reached, which is the same definition as for the enhancement device. In fact, all the equations for enhancement devices such as Eqs. (5-12) and (5-17) can be used, except V_T is now of opposite algebraic sign as the enhancement device threshold.

A graphic way to describe depletion devices is shown in Fig. 5-22b. With constant drain voltage in the saturation region, current flows as a parabolic function of gate voltage minus threshold voltage. For depletion devices, since

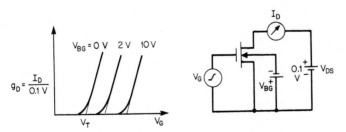

Figure 5-21 Setup and resulting plot for generating a conductance curve.

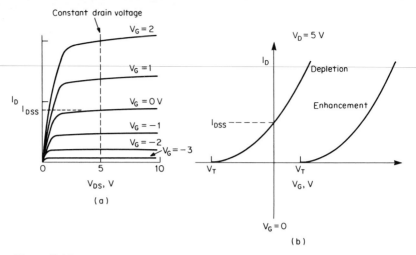

Figure 5-22 (*a*) **Transfer characteristics for a depletion-mode device, and** (*b*) **its transfer curve showing** I_D **vs.** V_G **for a constant drain voltage.**

V_T is to the left of the origin, then at $V_G = 0$, a certain current is already flowing. This current is called I_{DSS} and is given by

(5-32) $$I_{DSS} = \tfrac{1}{2}\beta(V_T)^2$$

Depletion transistors are extensively used in NMOS technology. To fabricate these devices, donors are ion-implanted with an energy that is designed to place the layer of charge very close to the surface. If phosphorus is used, it can be implanted through the gate oxide. With arsenic, the energy required to penetrate the oxide would be too high, so the implantation is performed before gate oxidation. In both cases, the implanted impurity concentration has to exceed the background *p*-type impurity concentration in order to form the built-in channel. This results in an *np* junction under the gate. The effect of the additional junction can be explained with the aid of the *CV* plots in Fig. 5-23. Note first that the gate capacitance is now the series combination of the oxide capacitance, the capacitance of the depletion layer extending from the surface into the *n* layer, *and* the junction capacitance of the *np* junction.

The explanation should best proceed from the C_{min} to C_{max} direction. At large positive gate voltages, which correspond to the inversion condition for normal *CV* structures, the semiconductor surface is accumulated (full of electrons) and the *n*-layer depletion capacitance is shorted out. This leaves the series combination of oxide capacitance and the junction capacitance for an effective capacitance that is slightly lower than the normal C_{min}. As the gate voltage is decreased, the surface region progresses from accumulation, through flatband, to depletion. The *n*-layer depletion capacitance is now inserted and the total series capacitance drops. As the gate voltage continues negative, the surface depletion layer expands until it just touches the edge of the np junction depletion region. The n layer is now fully depleted and the built-in channel is pinched off. This, by definition, is V_T.

For gate voltages more negative than V_T, which corresponds to the depletion condition for normal CV structures, the np structure starts to forward-bias, resulting in an increase in its capacitance as well as for the overall device. The total capacitance is thus at a minimum at V_T.

At extreme negative voltages, which corresponds to the accumulation condition for normal CV structures, the surface n layer actually inverts to p type, shorting out the surface depletion layer and the forward-biased np junction, leaving only the oxide capacitance. The total capacitance then goes to a maximum.

Stating the previous discussion in another way: because the surface is now of opposite type to the substrate, then its surface band bending (i.e., accumulation, depletion, or inversion) is the exact opposite of the case for the homogeneous substrate.

If the implanted impurity concentration is less than the background p-type impurity, then the substrate remains all p-type and the dip in capacitance is not manifested. Instead, a simple parallel shift of the CV curve takes place.

MOSFET Connected as Load Devices

Depletion-mode devices are most often used as load devices. To appreciate the unique properties of depletion-mode transistors as load devices, it is appropriate to consider in general the use of transistors as loads.

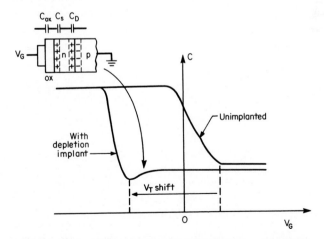

Figure 5-23 Comparison of CV plots between unimplanted enhancement-mode and implanted depletion-mode n-channel devices.

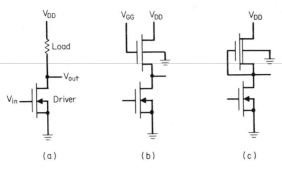

Figure 5-24   Use of resistor, enhancement- and depletion-mode load devices in a MOS inverter.

Various implementations of a MOS inverter are shown in Fig. 5-24. The inverter forms the basic building block of all MOS logic circuits. It consists of a driver device and load device. The high packing density of MOS integrated circuits as compared to bipolar ICs is due largely to the fact that the load device can be a MOS transistor which takes up less area than a resistor. The load device can be either an enhancement- or a depletion-mode transistor.

1. *Saturated Enhancement Load*   In Fig. 5-24*b*, both devices are enhancement type, and with the gate of the load device tied to its drain ($V_{GG} = V_{DD}$), the load device is always in saturation. This form of load device occupies a small amount of area since the drain is right beside the gate. However, the output voltage can only swing up to a level that is one threshold drop ($V_T + \Delta V_T$) below V_{DD}. Any attempt to move the output node past this point toward V_{DD} will shut off the load device. In addition, the charge-up time is quite long because the load device is losing current drive as the output charges up.

2. *Triode Enhancement Load*   If the gate of the load device is tied to a separate power supply more positive (for *n*-channel devices) than V_{DD} such that $V_{GG} - V_T > V_{DD}$, then the load device will always remain in the triode region and the output node can charge up to V_{DD}. There is some gain in speed because the load device is biased-on harder during the charge-up time. The disadvantages of this solution are the obvious need for an extra power supply and the extra chip area used up in routing this power supply lead to all the inverters. This solution is thus only used selectively in situations where the wider voltage swing and gain in speed is necessary, such as to serve clock-line drivers.

3. *Depletion Load*   One configuration that combines the advantages of the previous two, while adding further improvement in performance, is the use of depletion devices as load. Figure 5-24*c* shows one such inverter

with the gate of the load tied to its source. The area occupied is again small because the gate is right next to the source. The output voltage can swing from close to ground to V_{DD}. There is also a large increase in speed, as can be seen from drawing the load lines of all the configurations onto Fig. 5-25. (A load line traces the locus of the *driver* transistor drain current and voltage, as constrained by the load *VI* characteristics and the power supply level.) The load lines in the figure are drawn to the same normalized maximum current. As the output voltage V_D rises from zero toward V_{DD}, the available current for charging the output node drops for most load devices. Only the ideal depletion device behaves more like an ideal current source and has maximum current flowing throughout most of the output voltage swing. Even if the presence of back bias is taken into account, more current than the other configurations is available for charging the output node. The net result is faster switching transitions.

MOSFETS as Resistors

If V_{DS} is kept small, the drain and source act like the two terminals of a resistor whose resistance is

$$(5\text{-}33) \quad R = \frac{1}{\beta(V_G - V_T)}$$

The device can be used as a variable resistor with the gate as the controlling (but isolated) terminal. This has been successfully used in automatic gain control (AGC) circuits.

The device can also be used as a switch, again with the gate as the controlling terminal. With several such transistors connected with a common source and their gates sequentially pulsed on and off, a multiplexing circuit can be built to select one out of many incoming signals. One advantage of

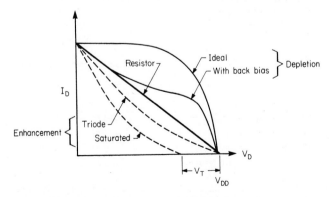

Figure 5-25 **Load lines for various load devices.**

using MOSFETs instead of bipolar transistors as switches is the absence of any offset voltage since the MOSFETs remain resistive down to $V_{DS} = 0$. Another advantage of MOSFETs is the high degree of isolation between the signal path and the gate electrode. With bipolar transistors, some current flow between the two is necessary to turn on the emitter-base junction.

Static Protection

The input of a MOS device is principally capacitive. Since that capacitance is extremely small, typically $\leq 10^{-12}$ pF, it represents an extremely high input impedance. Any small static charge that may be induced on the input leads not only has no way of leaking off but would induce a rather large voltage. If this voltage causes the gate oxide to exceed its critical breakdown field, the oxide would be irreversibly damaged. The breakdown field is normally 10^7 V/cm, or 100 V for 1000-Å oxide, but normal distribution in processing can cause some devices to break down at voltages two to three times lower than this value.

Static electricity is often generated by the human body (insulated from an earth ground by shoes) and transmitted to the device during handling. The voltage generated is highest when the humidity is low and can range up to tens of thousands of volts as Table 5-1 shows.[6] The solution to the static discharge problem is two-pronged. One is to prevent static buildup or discharge during manufacturing and up to the point of use. The second is to design input protection devices on the chip to make the circuit immune to normal occurrences of static discharge.

Static discharge during manufacture often occurs only after wafer fabrication because that is where the bulk of the direct human contact takes place and because leads are then individually accessible. But the same care should be taken in wafer fabrication to avoid static discharge since the prob-

TABLE 5-1 **Typical Electrostatic Voltage Levels Found on Personnel Engaged in Various Activities**

Activity	Voltage	
	10 to 20% humidity, V	65 to 90% humidity, V
Walking on carpet	35,000	1,500
Walking on vinyl floor	12,000	250
Working at bench	6,000	100
Handling vinyl envelopes	7,000	600
Picking up polyethylene bag	20,000	1,200
Sliding on foam-padded chair	18,000	1,500

Source: After Ref. 6.

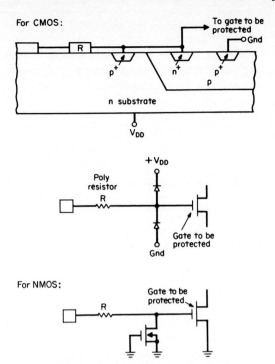

Figure 5-26   Input protection circuitry against static electricity discharge for CMOS and NMOS.

lem can show up anywhere. There have been instances reported, for example, of high-pressure jets of high-resistivity (ultrapure) deionized water causing oxide breakdown. This takes place during wafer scrubbing operations that follow the gate cycle. To avoid static discharge in situations where human handling is involved, the safest approach is to provide discharge paths for static electricity generated in the area. It may take the form of grounding all work tables and tools, and using grounded wrist straps, conductive benches, and floor mats. Raising the humidity or blowing in ionized air is also effective. For shipping, all packing materials such as trays, boxes, pouches, or tubes should not be capable of generating electricity. An even better procedure is to ship the ICs plugged into conductive foam so that all leads are shorted together until point of use.

It is now a universal practice to add static discharge protection circuitry to all inputs. There are many such circuits in use, each with varying degrees of effectiveness. Figure 5-26 shows two circuits that demonstrate the key elements of all of them. In the CMOS circuit, negative voltage spikes are clamped by the turn-on of the diode connected to ground, while positive voltage spikes are clamped by the upper diode to one diode drop above V_{DD}.

The resistor is necessary to limit current flow and prevent the diodes from being destroyed. Otherwise the gate would be unprotected from subsequent discharges. For NMOS circuits, two independently isolated diodes are not available; instead, the punch-through voltage of an enhancement device (either active or field device) is used. The punch-through voltage is much lower than the reverse breakdown of a simple diode but is still high enough to allow proper gate voltage swings. The punch-through mode is assured by grounding the gate. For negative voltage spikes, the drain junction of the punch-through device will forward-bias. If an even lower protection voltage is necessary, a zener breakdown between two heavily doped diffusions can be used. In all cases, a resistor to limit the current is needed. If the resistance is too large for fast logic circuits, the resistor can be eliminated while the punch-through device is replaced by a distributed version with series resistance in its drain. The drain also has a large junction periphery to facilitate early diode breakdown.

Output leads normally do not require protection circuits because they are tied to drain junctions which act as natural protection circuits. However, for extra protection, large periphery diodes may be added in parallel, or the drain junction may be surrounded by ground contacts to reduce voltage drops when static discharge current does flow.

In order to test the effectiveness of static protection circuitry, a circuit such as in Fig. 5-27 can be used (Military Spec. #883B). The 100-pF capacitor simulates the capacitance of a person and is charged to the high-voltage power supply. The relay then switches to discharge the capacitor to the item under test. The 1.5 kΩ series resistor represents the series resistance of the discharge path through the human body (the resistance is primarily that of the skin). All components of the test circuit must be rated for the highest voltage used. In particular, the relay should be a high-voltage reed or mercury-wetted relay to avoid contact bounce. The whole test setup should be in a safety enclosure with the high-voltage sections shorted out the moment the enclosure is opened. In performing the tests, one should judge the effectiveness of the protection circuits with not just one discharge to a given lead, but rather several discharges, such as 5 to 10. And the failure criteria should be not just for destruction of the input but also for significant degradation of the overall device performance. In practice the most immediate degradation observed is usually an increase in input leakage current.

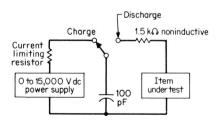

Figure 5-27 Circuit to test the effectiveness of static protection circuitry. (*After Ref. 6 and military spec., #883B.*)

REFERENCES

1. P. Richman, *Characteristics and Operation of MOS Field-Effect Devices,* Chap. 4, McGraw-Hill, New York, 1967.

2. R. M. Swanson and J. D. Meindl, "Ion-implanted complementary MOS transistors in low voltage circuits," *IEEE J. Solid-State Circuits,* vol. SC-7, no. 2, pp. 146–153, April 1972.

3. J. A. Cooper, Jr., "Limitations on the performance of field-effect devices for logic applications," *Proc. IEEE,* vol. 69, no. 2, pp. 226–231, February 1981.

4. R. S. Muller and T. I. Kamins, *Device Electronics for Integrated Circuits,* Chap. 8, Wiley, New York, 1977.

5. T. W. Sigmon and R. Swanson, "MOS threshold shifting by ion implantation," *Solid-State Electronics,* vol. 16, pp. 1217–1232, 1973.

6. R. M. Rzepecki, "Static protection for electronic components," *Machine Design,* pp. 97–101, March 26, 1981.

7. J. R. Brews, "Physics of the MOS transistor," in *Silicon Integrated System, Part A,* D. Kahng, Ed., pp. 1–118, Academic Press, New York, 1981.

8. E. H. Nicollian and J. R. Brews, *MOS Physics and Technology,* Wiley, New York, 1982.

PROBLEMS

1. Derive the triode device equation for a circular MOSFET whose cross section is shown in Fig. 5-28. *Hint: W* is no longer a constant.

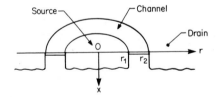

Figure 5-28 Problem 1.

2. For the 10–40 method of measuring threshold, show that

$$V_T = 2V_{10} - V_{40}$$

3. Figure 5-29 shows the two-terminal characteristics of a p-channel device whose gate is tied to its drain. Use the square root of I_D method to measure its V_T and β. Use six to eight points for your graph. Compare the value of V_T obtained with that from the 10–40 method.

4. In more recent MOS processes, there is a trend toward using a lightly doped substrate and a very shallow ion implantation to adjust the threshold to any desired

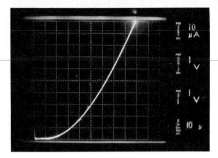

Figure 5-29 Problem 3.

value. What would be the effect on the change in threshold due to back-gate bias? The speed of the whole circuit is also improved. Why?

5. If the depletion implant is too deep, then the built-in channel could not be shut off even with an accumulating gate voltage. Make a plot of what you expect the conductance plot to look like.

6. Derive the triode equation with varying bulk charge, i.e., Eq. (5-25).

6

Small-Geometry Effects

PROFESSOR LEX AKERS

Arizona State University

6-1 INTRODUCTION

As MOS devices are scaled down to near and submicrometer dimensions, geometry effects resulting from this scaling can produce variations in device performance. Such variations can have a severe impact on circuit design and performance and therefore must be predictable.

Geometry effects include the variations in the threshold voltage and subthreshold current as a function of short, narrow, and small device dimensions. An accurate prediction of the threshold voltage is needed to determine circuit noise margins, speed, and required node voltages. An accurate prediction of the subthreshold current is needed to determine off-state power dissipation and memory refresh times. Other important effects that occur as devices become smaller are velocity saturation, hot electron injection, and enhanced device performance as devices shrink.

This chapter provides a discussion of these topics. Since much of the research in these areas is currently still under way, only the fundamental concepts will be discussed. The interested reader will have to go to the current literature to obtain the latest information on this dynamic field.

6-2 NONUNIFORM DOPING AND EFFECT ON THRESHOLD VOLTAGE

In the theory and examples used in the text so far, we have assumed uniform substrate doping. The reason for this assumption is straightforward; it greatly simplifies the mathematics. But in a real device, the doping profile is far from uniform in most structures. Nonuniform doping is generally found because of the redistribution of impurity ions during thermal oxidation. Besides the nonuniformities in doping generated by high-temperature processes during fabrication, often various nonuniform doping profiles are specifically caused by ion implantation to alter and improve a device characteristic. For example, with ion implantation, one can fabricate self-aligned sources and drains that are shallow, have very little lateral diffusion, and therefore have a small source or drain to gate capacitance while simultaneously doping the polysilicon gate.

A nonuniformly doped substrate will have a definite effect on the threshold voltage. The depth of the nonuniform implant as well as the dose will explicitly modify the standard threshold voltage expression. In most cases, the true doping profile does not allow closed-form threshold voltage expressions, so approximations to its shape are used. However, these approximations are very helpful in obtaining the trends in the shift in the threshold voltages for various implant profiles.

One commonly used approximation of a doping profile is the step-doping profile,[1] as shown in Fig. 6-1. After annealing, the real implant profile will in many cases approach this structure. The profile is modeled with a surface concentration of N_S with a depth of X_D. At X_D the concentration will abruptly change to the substrate doping level N_A.

The surface doping concentration is found from the real doping profile by averaging over X_D:

$$(6\text{-}1) \quad N_S = \frac{1}{X_D} \int_0^{X_D} N(x) \, dx$$

The dose D_I is

$$(6\text{-}2) \quad D_I = (N_S - N_A)X_D$$

In many devices, a very shallow implant is used to modify the threshold voltage. The limiting case would be an infinitely thin sheet,[2] approximately a delta function, of ionized charge localized at the SiO_2–Si interface. This

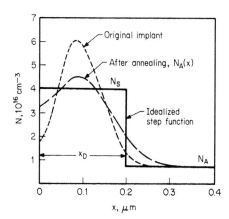

Figure 6-1 Doping profile under gate oxide. (*After Ref. 1.*)

is equivalent to modifying the fixed interface oxide charge and hence changing the flatband voltage. For ionized boron atoms, the flatband voltage will be effectively reduced and the threshold voltage will be increased. The flatband voltage is changed to

(6-3) $\quad V_{FB} \text{ (implant)} = V_{FB} \text{ (no implant)} + \dfrac{qD_I}{C_o}$

Therefore the original threshold voltage expression is modified to be

(6-4) $\quad V_T = V_{FB} + 2\phi_f + \dfrac{\sqrt{2\epsilon_s\epsilon_o qN_A(2\phi_f + V_{BG})}}{C_o} + \dfrac{qD_I}{C_o}$

and the shift in the threshold voltage is

(6-5) $\quad \Delta V_T = V_T \text{ (implant)} - V_T \text{ (no implant)} = \dfrac{qD_I}{C_o}$

The calculation can become more complex as X_D becomes deeper. If the depletion depth is totally contained in the implanted region $X_D \geq W_D$, where W_D is the depletion depth, then the surface can be considered to be uniformly doped at N_S. The standard V_T expression can be used with N_A replaced with N_S and ϕ_f calculated using N_S.

If $W_D > X_D$, Poisson's equation must be solved using a high-low step-doping profile instead of a uniform profile. Performing this integrating gives a depletion depth of[3]

(6-6) $\quad W_D = \sqrt{\dfrac{2\epsilon_s\epsilon_o}{qN_A}\left[\phi_s + V_{BG} - \dfrac{qX_D^2}{2\epsilon_s\epsilon_o}(N_S - N_A)\right]^{1/2}}$

This gives a threshold voltage of

(6-7) $\quad V_T = V_{FB} + \phi_s + \dfrac{\sqrt{2q\epsilon_s\epsilon_o N_A}}{C_o}\left(\phi_s + V_{BG} - \dfrac{qX_D}{2\epsilon_s\epsilon_o}D_I\right)^{1/2} + \dfrac{qD_I}{C_o}$

Figure 6-2 shows the threshold voltage vs. back-gate bias for various doping profiles. Notice for shallow X_D, i.e., $X_D \ll W_D$, the threshold voltage approaches the delta function dose. For $X_D > W_D$, the initial threshold voltage is equivalent to a uniformly N_S doped region. As V_{BG} is increased, W_D approaches and soon surpasses X_D. The threshold voltage will vary from the uniform doped N_S value to the implant appearing to reside at the surface.

6-3 SUBTHRESHOLD CURRENT

Classical theory sets the channel current equal to zero when $V_G = V_T$. In a real device, this is not the case. A nonzero current occurs when the surface

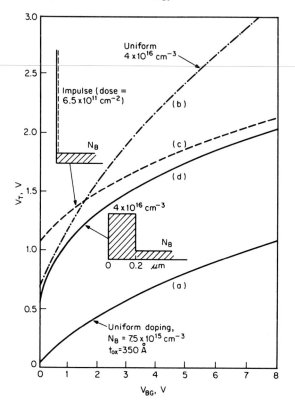

Figure 6-2 Threshold voltage vs. V_{BG} for various doping profiles. *(After Ref. 1.)*

is in the weak inversion region, i.e., $\phi_f \leq \phi_s \leq 2\phi_f$. This corresponding current is defined as the subthreshold current.[4] Subthreshold current is important in that it is a leakage current that will discharge capacitors and waste power.

In weak inversion for long channel devices, the surface potential is constant from source to drain, and therefore the longitudinal channel electric field is zero. Hence, the only current that can flow is diffusion current. This condition is analogous to current flow in a bipolar junction transistor, where the source is the emitter, the channel is the base, and the drain is the collector. The diffusion current is

$$(6\text{-}8) \quad I = qD_nA\frac{\partial n}{\partial x}$$

where A is the cross-sectional area, and n is the electron density.

Current continuity in steady state without generation or recombination requires

(6-9) $\quad \dfrac{\partial J}{\partial x} = 0$

and hence this forces the electron density to be linear with distance. Therefore,

(6-10) $\quad n(x) = - \left(\dfrac{n(0) - n(L)}{L} \right) x + n(0)$

The subthreshold current is then

(6-11) $\quad I_D = qAD_n \left[\dfrac{n(0) - n(L)}{L} \right]$

The electron densities at the source and drain are, respectively,

(6-12) $\quad n(0) = n_i \exp \left[B(\phi_s - \phi_f) \right]$

(6-13) $\quad n(L) = n_i \exp \left[B(\phi_s - \phi_f - V_D) \right]$

where ϕ_s = surface potential
$B = q/kT$
V_D = drain voltage
ϕ_f = bulk potential defined as

(6-14) $\quad \phi_f = \dfrac{kT}{q} \ln \left(\dfrac{N_A}{n_i} \right)$

The area in which the current flows is the channel width times the effective channel depth. The exact depth of the channel is difficult to calculate. It can be approximated as the depth the potential decreases by kT/q. Therefore, the inversion thickness is $kT/(q \cdot E_s)$, where E_s is the electric field at the surface and is given by[3]

(6-15) $\quad E_s = - \dfrac{Q_B}{\epsilon_s \epsilon_o} = \sqrt{\dfrac{2qN_A\phi_s}{\epsilon_s \epsilon_o}}$

Substituting Eq. (6-15) into Eq. (6-11) gives[5]

(6-16) $\quad I_D = q\mu \dfrac{Wn_i \exp \left[B(\phi_s - \phi_f) \right]}{LB\sqrt{\phi_s}} \left[1 - \exp (BV_D) \right]$

It is important to notice that I_D is exponentially related to ϕ_s, and for $V_D \geq 100$ mV, the subthreshold current is not a function of drain voltage for long channels.

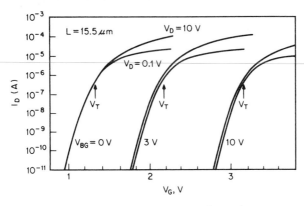

Figure 6-3 Subthreshold current plot. *(After Ref. 11.)*

Figure 6-3 is a plot of the subthreshold current. Notice as V_G approaches the threshold voltage, the slope goes from linear (on a log plot) to sublinear (linear on a linear plot).

6-4 SHORT-CHANNEL EFFECT

A MOS device is considered short when the channel length is the same order of magnitude as the source and drain junction depletion depths.[6] Modulation of the threshold voltage for short-channel devices has been observed experimentally but is not predicted by the classical threshold voltage expression. This modulation results in a reduction of threshold voltage as the channel length becomes small and is one of several of the short-channel effects. Before discussing the various models describing the short-channel effect, it is relevant to review the classical threshold voltage derivation.

The classical threshold voltage expression is derived by assuming that both the length and width of the devices are large. Thus we may assume that the electric field normal to the gate electrode is much greater than the electric field parallel with the channel. Hence a one-dimensional treatment of the problem is sufficient. It is important to note that this analysis implies that the built-in potentials produced by both the source- and drain-depletion regions can be neglected. This is not a valid assumption for short-channel devices.

The threshold voltage expression for a large-geometry MOSFET can be obtained from the charge conservation law in the region bounded by the gate electrode and the semiconductor bulk. The sum of this charge is

(6-17) $$Q_G + Q_f - (Q_n + Q_B) = 0$$

where Q_G = charge on polysilicon gate electrode
  Q_f = fixed charge in SiO_2

Q_n = inversion-layer charge due to free carriers induced in channel region

Q_B = bounded charge per unit area due to the ionized impurity concentration in the depletion region

The definition of the threshold voltage implies $Q_n \cong 0$. Expressing Eq. (6-17) in terms of voltages for an n-channel MOSFET gives

(6-18)
$$V_G = V_{\text{FB}} + \phi_s + \frac{Q_B}{C_o}$$

where V_{FB} = flatband voltage, including effects of fixed surface charge, distributed charge in oxide, and difference in work functions

ϕ_s = surface potential

C_o = oxide capacitance per unit area

At turn-on, the surface potential locks at approximately $2\phi_f$. Therefore, the threshold voltage for a large-geometry MOSFET is given by

(6-19)
$$V_T = V_{\text{FB}} + 2\phi_f + \frac{Q_B}{C_o}$$

where $Q_B = q N_A W_c$

q = electric charge

N_A = effective doping concentration

W_c = channel-depletion depth

Equation (6-19) is valid as long as the channel length is long compared to the source and drain junction depletion depths and the width is wide compared to the depth of the gate-induced depletion region.

Figure 6-4 illustrates the cross section of a large-geometry MOSFET. The depth of the source- and drain-depletion regions W_S and W_D can be

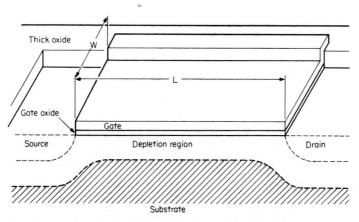

Figure 6-4 Cross section of large geometry MOSFET. *(After Ref. 6.)*

obtained from Poisson's equation using the abrupt junction approximation. This is a good assumption since both source and drain regions are doped much higher than the substrate. The depletion approximation[7] can be applied under the gate in the substrate to obtain the channel depletion depth W_c.

The concept of charge sharing both provides physical insight into the development of the threshold voltage expression and predicts trends correctly. From Eq. (6-19) the total amount of charge under the gate determines in part the threshold voltage. If the channel length L is much greater than either the source- or drain-depletion regions, the charge sharing which occurs from these regions can be neglected. Using this assumption and Fig. 6-4, the total amount of charge induced by the gate voltage is approximately given by

$$(6\text{-}20) \quad Q_{BT} \approx q N_A W_c W L$$

or

$$(6\text{-}21) \quad Q_B = \frac{q N_A W_c W L}{W L} = q N_A W_c$$

where Q_{BT} is the total bound charge, and W is the width of the device. Hence Eq. (6-19) can be considered a very good approximation for the threshold voltage in large devices.

If the length is reduced to a distance comparable to W_S and W_D, part of the charge which is induced under the gate is also contained in the source or drain regions (Fig. 6-5). In other words, charge sharing is occurring in both the source- and drain-depletion regions with the gate region. This implies that the total effective charge induced under the gate can no longer be approximated by a rectangular region. Hence the amount of charge

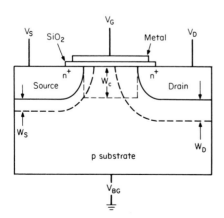

Figure 6-5   **Charge-sharing model.**

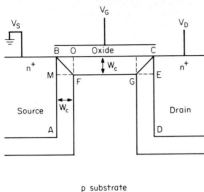

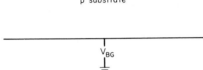

p substrate

V_{BG}

Figure 6-6 Approximate short-channel model. (*After Ref. 8.*)

reflected to the gate electrode can be expected to be reduced. Since Q_B reflects this charge reduction as L decreases, the threshold voltage given by Eq. (6-19) will also decrease.

A two-dimensional model[8] which demonstrates the short-channel effect and also allows a closed-form solution of the threshold voltage is shown in Fig. 6-6. Though simple, this model provides physical insight into the charge-sharing concept. The area outlined by $BCEM$ represents the rectangular region containing the total charge induced under the gate for a long-channel device with the substrate grounded. As the channel length is reduced, the effects of the source- and drain-depletion regions cannot be neglected. For the case when both source and drain are grounded and the surface is strongly inverted, the potential along $ABCD$ is zero. Hence, the potential of MBO is zero. The maximum potential from F to M is approximately equal to the maximum potential from F to O. This is true since at inversion the built-in potential V_{Bi} is to a rough approximation equal to $2\phi_f$. Thus we may assume $W_S = W_c$. This implies that the charge contained in the square $FMBO$ is equally supported by both source and gate. In other words, half the charge can be considered to be supported by the gate while half is supported by the source. Assuming $V_D \cong 0$, the same analysis can be applied to the drain region. Hence the total charge enclosed can be approximated by the trapezoidal region $BCGF$. This simple model becomes more accurate with an applied back-gate bias since the maximum potentials at FM and FO correspond to $V_{BG} + V_{Bi}$ and $V_{BG} + 2\phi_f$ respectively at inversion. By subtracting the triangular regions donated by FMB and CEG from the rectangular region $MBCE$, the total volume compromising the modified bulk charge term is[8]

$$(6\text{-}22) \quad Q_{BT} = qN_A(WLW_c) - 2W\left(\frac{W_c W_c}{2}\right)$$

Using Equation (6-22), the threshold voltage is

$$(6\text{-}23) \quad V_T = V_{FB} + 2\phi_f + \frac{qN_A(WLW_c - WW_c^2)}{C_o WL}$$

or

$$(6\text{-}24) \quad V_T = V_{FB} + 2\phi_f + \frac{Q_B}{C_o}\left(1 - \frac{W_c}{L}\right)$$

with

$$W_c = \left[\frac{2\epsilon_s\epsilon_o}{qN_A}(V_{BG} + 2\phi_f)\right]^{1/2}$$

where $2\phi_f$ is the surface potential at the onset of strong inversion, and $\epsilon_s\epsilon_o$ is the dielectric constant of the semiconductor.

A model[9] which is a more accurate representation of the structure of a MOS device is shown in Fig. 6-7. Notice that both the source and drain junctions are characterized by a radius of curvature r_j. This is a much better approximation of the shape of the diffused junction profiles. By making the assumptions that $W_S = W_D = W_c$ for $V_D \cong 0$, we are able to observe the trapezoidal region of charge supported by the gate. Notice that this trapezoidal region arises from charge sharing which is occurring in both the source- and drain-depletion regions with the gate region. This leads to the bulk charge being modified in the threshold voltage expression given in Eq. (6-19).

In order to obtain the threshold voltage expression, the surface potential must be defined throughout the channel at inversion. Unfortunately any MOS device which is modeled by using the trapezoidal approximation does

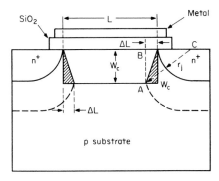

Figure 6-7 **Short-channel model with curved source-drain junctions.** (*After Ref. 9.*)

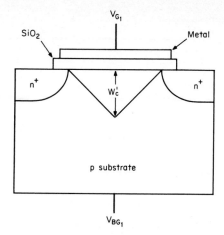

Figure 6-8 Short-channel model when depletion regions touch. (*After Ref. 9.*)

not have a constant surface potential along the channel. To circumvent this difficulty, an arbitrary smoothing approximation such that the effect of the bulk charge on the surface potential is uniform along the channel at inversion is made. Hence the threshold voltage expression may be derived. By subtracting the shaded triangular regions or charge associated with both the drain and source region from the rectangular region of charge, we obtain the total volume comprising the modified bulk charge[9]

$$(6\text{-}25) \quad Q_{BT} = qN_A\left(WLW_c - 2W\frac{\Delta LW_c}{2}\right)$$

where ΔL can be obtained by using the triangle encompassed by *ABC*. Hence

$$(6\text{-}26) \quad (\Delta L + r_j)^2 + W_c^2 = (W_c + r_j)^2$$

Using the above results the threshold voltage expression obtained is

$$(6\text{-}27) \quad V_T = V_{\text{FB}} + 2\phi_f + \frac{Q_B}{C_o}\left\{1 - \left[\left(1 + \frac{2W_c}{r_j}\right)^{1/2} - 1\right]\frac{r_j}{L}\right\}$$

An interesting result is that the threshold voltage is a function of junction radius of curvature r_j. This was not included in the previous model due to the deep rectangular junction profiles assumed. This same approach can be applied to nonuniform doping in the bulk by integrating the charge density over the area enclosed under the gate. Poisson's equation must be solved for the case of nonuniform doping to obtain the channel depletion width W_c.

Equation (6-27) is valid as long as the back-gate bias voltage V_{BG} is less than the voltage V_{BG1} needed to cause the source- and drain-depletion regions to meet. As the back-gate bias voltage is increased to the point where both regions touch, the charge enclosed is given by the triangular shaped region shown in Fig. 6-8. If W_c' is denoted as the channel depth when both the source and drain regions meet, then

$$(6\text{-}28) \quad W'_c = \frac{L}{2r_j}\left(r_j + \frac{L}{4}\right)$$

Using Eqs. (6-27) and (6-28), V_{BG1} can be determined.

Effects of Drain Voltage on Threshold Voltage

The model used to describe the MOSFET shown in Fig. 6-7 assumes the drain is shorted to the source. This configuration is of course not true in any realistic device structure. Therefore, we must modify the short-channel threshold voltage expression to include the drain voltage V_D.

The most immediate effect of the drain voltage is the increase in the extent of the drain-depletion region into the channel. The extension of the depletion region into the channel can be approximated by the abrupt diode depletion depth as

$$(6\text{-}29a) \quad \sqrt{\frac{2\epsilon_s\epsilon_o(V_{Bi} + V_{BG})}{qN_A}}$$

where V_{BG} is the back-gate bias, and V_{Bi} is the built-in potential.

For zero back gate, the extent of the drain-depletion region is from 0.3 to 0.8 μm for most typical devices. If a drain voltage of 5 V is applied, the

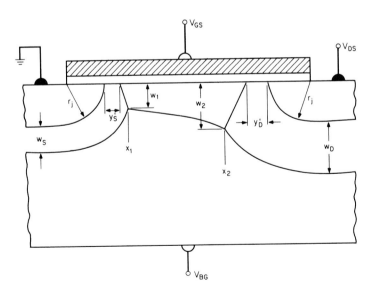

Figure 6-9   Short-channel model for $V_D > 0$.

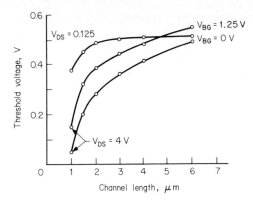

Figure 6-10 Threshold voltage for various V_D.
(*After Ref. 5.*)

expression to calculate the extension of the depletion region is modified to be

$$(6\text{-}29b) \qquad \sqrt{\frac{2\epsilon_o\epsilon_s(V_{\text{Bi}} + V_{\text{BG}} + V_D)}{qN_A}}$$

For a drain voltage of 5 V, the depletion region can expand to greater than 2 μm. This increase is not negligible and must be accounted for in our threshold voltage expression. Using the geometrical approximation shown in Fig. 6-9, the threshold voltage can be derived in a similar manner as Eq. (6-27) to be[10]

$$(6\text{-}30) \qquad V_T = V_{\text{FB}} + 2\phi_f + \frac{\sqrt{2q\epsilon_s\epsilon_o N_A(\phi_s + V_{\text{BG}})}}{C_o}$$

$$\times \left[1 + \frac{\left(L - \sqrt{\dfrac{2\epsilon_s\epsilon_o}{qN_A}(V_{\text{Bi}} + V_{DS} + V_{\text{BG}})} - \sqrt{\dfrac{2\epsilon_s\epsilon_o}{qN_A}(V_{\text{Bi}} + V_{\text{BG}})}\right)}{\left(L - \sqrt{\dfrac{2\epsilon_s\epsilon_o}{qN_A}(V_{\text{Bi}} - 2\phi_f)} - \sqrt{\dfrac{2\epsilon_s\epsilon_o}{qN_A}(V_{\text{Bi}} + V_{DS} - 2\phi_f)}\right)} \right]$$

Figure 6-10 shows the effect of V_{DS}.

Subthreshold Current in Short-Channel Devices

In a long-channel MOSFET, the drain voltage has a negligible effect on the fields in the source region. As the channel length is reduced, this is no longer the case as the drain field extends into the source region.[11] This modulation of the source potential has a direct and amplifying effect on the subthreshold current. To include the short-channel case in the original sub-

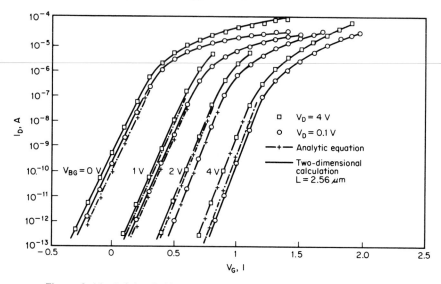

Figure 6-11 Subthreshold current for a short-channel MOSFET. *(After Ref. 5.)*

threshold current expression Eq. (6-16), one can replace the channel length by an effective channel length L_{eff}. The effective channel length is defined as[10]

$$L_{\text{eff}} = L - y_S - y_D$$

where

(6-31) $$y_S = \sqrt{\frac{2\epsilon_s\epsilon_o}{qN_A}(V_{\text{Bi}} - 2\phi_f)}$$

and

(6-32) $$y_D = \sqrt{\frac{2\epsilon_s\epsilon_o}{qN_A}(V_{\text{Bi}} - 2\phi_f + V_{DS})}$$

While the subthreshold current in a long device is independent of the drain voltage for $V_D \geq 100$ mV, for a short-channel device the current increases with V_D (see Fig. 6-11). Also, as devices are made with smaller channel lengths, the magnitude of the current increases and the slope of the subthreshold current is reduced. This increase in current can cause a significant leakage problem in small device circuits.

6-5   NARROW-WIDTH EFFECT

Device width has been found to have a significant effect on device behavior.[12] A MOSFET is considered narrow if the width of the device is the same order of magnitude as the depth of the channel-depletion region. In practice, for a channel-depletion depth of 0.5 μm, device widths of 5 μm or less affect the electrical behavior of the device.

Figure 6-12 illustrates a simplified width cross section of a MOSFET. The polysilicon gate overlaps the nonrecessed thick oxide on both sides of the thin gate oxide.

The narrow-width effect is characterized by an increase in the threshold voltage as the width W is reduced. This results from the charge stored under the thick oxide region. As W is reduced, the volume of charge in the gate region is reduced, whereas in the thick oxide region the volume of charge remains constant. This fixed amount of charge becomes increasingly significant as W is reduced and contributes to an increased threshold voltage. In Fig. 6-13, a plot of V_T versus W for different doping concentrations demonstrates this behavior.

Any numerical expression to predict V_T for a narrow-width device must include the effect of the fixed side charge. The standard V_T expression, Eq. (6-19), only includes the charge in W_c and therefore is very inaccurate for small W.

A closed-form expression for the threshold voltage of a narrow-width MOSFET can be derived as follows. Consider three geometrical approximations of the area enclosing the fixed extra charge: a triangle, a quarter circle, and a square. For uniform substrate doping, the extra charge ΔQ_W enclosed in each area is[12]

$$
\begin{array}{ll}
(6\text{-}33) & \\
& \\
(6\text{-}34) \quad \Delta Q_W = \left\{
\begin{array}{ll}
\dfrac{qN_A W_c{}^2}{2} & \text{triangle} \\[3mm]
\dfrac{qN_A \pi W_c^2}{4} & \text{quarter circle} \\[3mm]
qN_A W_c^2 & \text{square}
\end{array}
\right. & \\
(6\text{-}35) & \\
\end{array}
$$

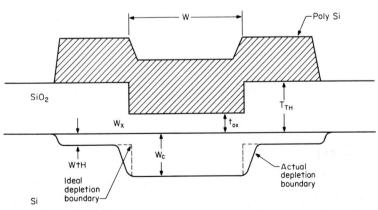

Figure 6-12 Width cross section. *(After Ref. 13.)*

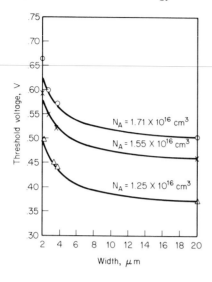

Figure 6-13 Threshold voltage vs. channel width for various doping levels. *(After L. A. Akers, Beguwala, and Custode, IEEE Trans. Electr. Dev., vol. ED-28, no. 12, p. 1490, 1981.).*

The charge on both sides of the device will contribute a voltage to the threshold voltage of

$$
(6\text{-}36) \qquad \frac{qN_A W_c^2}{C_o W} \qquad \text{triangle}
$$

$$
(6\text{-}37) \qquad V_{\Delta QW} = \begin{cases} \dfrac{qN_A \pi W_c^2}{2C_o W} & \text{quarter circle} \end{cases}
$$

$$
(6\text{-}38) \qquad \frac{2qN_A W_c^2}{C_o W} \qquad \text{square}
$$

Hence, a general form of the threshold voltage for a narrow-width MOSFET is

$$
(6\text{-}39) \qquad V_T = V_{\text{FB}} + 2\phi_B + \frac{qN_A}{C_o}\left(W_o + \frac{\delta W_c^2}{W}\right)
$$

where

$$
(6\text{-}40) \qquad \delta = \begin{cases} 1 & \text{triangle} \\ \tfrac{1}{2}\pi & \text{quarter circle} \\ 2 & \text{square} \end{cases}
$$

6-6 SMALL-GEOMETRY EFFECTS

A small-geometry MOSFET is defined as a device with a channel length the same order of magnitude as the junction-depletion depth, and the channel width the same order of magnitude as the channel-depletion depth. For such small devices, neither just the short-channel nor the narrow-width expression is sufficient to predict the threshold voltage accurately. An expression that includes both of these effects plus the coupling of these effects is required.

The narrow-width model of Sec. 6-5 can be extended to develop a threshold voltage expression for small MOSFETs. This expression is derived from a three-dimensional geometrical approximation of the bulk charge in the depletion region. The interdependence of the short-channel and the narrow-width effect is explicit.

The three-dimensional geometrical approximation of the bulk charge is obtained by summing a length cross-sectional area over the width direction. The length cross-sectional area is

$$(6\text{-}41) \quad \left[1 - \left(\sqrt{1 + \frac{2W_c}{r_j}} - 1 \right) \frac{r_j}{L} \right] LW_c$$

This length cross-sectional area is summed over the width direction where charge is stored, a distance of $W + 2W_c$.

The three-dimensional volume approximation of the depletion region is[13]

$$(6\text{-}42) \quad \left[1 - \left(\sqrt{1 + \frac{2W_c}{r_j}} - 1 \right) \left(\frac{r_j}{L} + \frac{2W_c r_j}{WL} \right) + \frac{2W_c}{W} \right]$$

Therefore, the threshold voltage is

$$(6\text{-}43) \quad V_T = V_{FB} + 2\phi_f + \frac{qN_A W_c}{C_o} \left[1 - \left(\sqrt{1 + \frac{2W_c}{r_j}} \right. \right.$$
$$\left. \left. - 1 \right) \left(\frac{r_j}{L} + \frac{2W_c r_j}{WL} \right) + \frac{2W_c}{W} \right]$$

Devices were fabricated and their experimental threshold voltages were determined using linear extrapolation to the gate voltage axis from a plot of channel current vs. gate voltage. The channel current vs. gate voltage was plotted on an *XY* recorder using 100-mV dc applied between the source and drain. Figure 6-14 shows the threshold voltage plotted vs channel length for varying channel width. The agreement between the small geometry threshold voltage expression and the data is close.

Figures 6-15 and 6-16 demonstrate the need for a small-geometry expression. Neither the short-channel nor the narrow-width expression will accurately predict the threshold voltage for a small-geometry structure.

6-7 SHRINK AND SCALING

Modern MOS ICs are quite complex and represent large investments in time, workforce, and resources to design. In order to decrease the cost of a

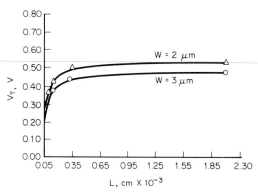

Figure 6-14   **Threshold voltage vs. channel length**
for various widths. $r_j = 0.4$ μm. *(After Ref. 13.)*

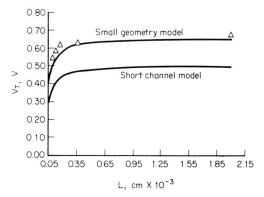

Figure 6-15   **Importance of narrow-width effect.** *W*
$= 2$ μm; $N_A = 1.71 \times 10^{16}$ cm^{-3}; $r_j = 0.2$ μm. *(After
Ref. 13.)*

circuit, the most effective route is to reduce the die size. However, it would
not be economically viable to redesign and re-lay out the chip; instead, the
mask set can simply be linearly shrunk. The shrinking process is equivalent
to an optical reduction, although in practice, it is accomplished by scaling
the digital information that is used to generate the mask. To illustrate the
power of this concept, consider that a 30 percent reduction in linear dimension results in twice as many available die locations. Chapter 14 shows that
the percent yield also goes up with reduction in chip size. The net increase
in good die is therefore even higher than the factor of two.

Perhaps almost as important, there is a net gain in performance with shrinking. First of all, there is no change in current driving capability of the transistors because the W/L ratio remains the same. However, the load capacitances the transistors drive are reduced dramatically. The gate capacitance, parasitic capacitance of metal, polysilicon, and diffusion interconnect lines are all reduced in proportion to the area reduction. The speed of the circuit therefore improves as a result.

The counterforce to a shrink is the fact that finer lines and spaces need to be resolved and etched. In practice, this limits the amount of shrink in the linear dimension to 5 to 10 percent each time. Indiscriminate shrinks to a circuit without changes in the process also result in devices showing short-channel effects. The solution is actually to scale down the devices. Scaling involves reducing vertical dimensions, such as oxide thicknesses and junction depths, as well as raising all doping concentrations. Increased capacitances negate some of the improvement in performance brought about by reduction in size. However, scaling does bring about improvement in device gain such that overall, shrinking with scaling will still result in significantly improved performance.

6-8 SCALING DOWN THE DIMENSIONS OF MOS DEVICES

The density and speed of MOS devices can be increased if the dimensions of the devices are scaled down. While there are a variety of methods[14] to scale devices, the simplest is constant electric field scaling.[15] Constant field scaling is when the dimensions of the device are scaled so that the electric field in the channel is held constant. This should minimize the high fields that cause mobility degradation and electron injection into the oxide.

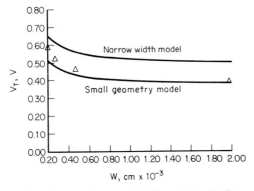

Figure 6-16 Importance of short-channel effect. *(After Ref. 13.)* $L = 1\ \mu m$; $N_A = 1.71 \times 10^{16}\ cm^{-3}$; $r_j = 0.2\ \mu m$. (After Ref. 13.)

If all the dimensions and voltages of a MOSFET are divided by a constant α, where α is greater than one, the device is altered in the following way:

$$L' = \frac{L}{\alpha}$$

$$W' = \frac{W}{\alpha}$$

$$t'_{ox} = \frac{t_{ox}}{\alpha}$$

$$V'_G = \frac{V_G}{\alpha}$$

$$V'_D = \frac{V_D}{\alpha}$$

$$N'_A = N_A \times \alpha$$

From this scaling we see the following effect on the transit time, gate capacitance, channel current, power dissipation per device and area:

$$\frac{\chi'}{\chi} = \frac{(L/\alpha)^2/(V/\alpha)}{(L^2/V)} \qquad \text{transit time}$$

Therefore

$$\chi' = \frac{\chi}{\alpha}$$

$$\frac{C'_o}{C_o} = \frac{(L/\alpha)(W/\alpha)/t_{ox}/\alpha}{LW/t_{ox}} \qquad \text{gate capacitance}$$

Therefore

$$C'_o = \frac{C_o}{\alpha}$$

$$\frac{I'}{I} = \frac{W/\alpha(V/\alpha)^2/(L/\alpha)(t_{ox}/\alpha)}{WV^2/Lt_{ox}} \qquad \text{current}$$

Therefore

$$I' = \frac{I}{\alpha}$$

$$\frac{P'_D}{P_D} = \frac{(I/\alpha)(V/\alpha)}{IV} \qquad \text{power dissipation per device}$$

Therefore

$$P'_D = \frac{P_D}{\alpha^2}$$

$$\frac{A'}{A} = \frac{(W/\alpha)(L/\alpha)}{WL} \qquad \text{area per device}$$

Therefore

$$A' = \frac{A}{\alpha^2}$$

Hence the results of scaling are increased device speed by α, density increase by α^2, and reduced power dissipation per device by α^2. With the doping level increased by α, the junction-depletion depths and the threshold voltage are reduced by approximately α. The problems with this scaling are that the subthreshold current remains essentially the same for both structures, the parasitic capacitance may not scale, and the interconnect resistance increases causing increased delay between devices. In fact for small devices, the interconnect delays could be the dominant delays and hence limit overall circuit performance. The effects of constant field scaling are shown below:

$$L' = \frac{L}{\alpha} \qquad V'_t \cong \frac{V_t}{\alpha}$$

$$W' = \frac{W}{\alpha} \qquad N'_A = N_A \alpha$$

$$t'_{\text{ox}} = \frac{t_{\text{ox}}}{\alpha} \qquad C'_o = \frac{C_o}{\alpha}$$

$$V'_D = \frac{V_D}{\alpha} \qquad R' = R\alpha$$

$$V'_G = \frac{V_G}{\alpha} \qquad \chi' = \frac{\chi}{\alpha}$$

6-9 SUMMARY

The effects of device geometry on the threshold voltage and the subthreshold current have been discussed. Also the effects of nonuniform doping and scaling on device performance have been described. Other factors affecting device and circuit performance as dimensions are reduced exist.

If the supply voltage is not scaled, as in many circuits, to stay compatible with TTL (transitor-transistor logic) voltage levels, internal electric fields can become very large. These large electric fields can cause electrons to be injected into the gate oxide resulting in gate current, threshold voltage shifts, and degraded device performance. Also, for high drain electric fields the velocity of the carriers will saturate and the device's current drive will be reduced. Closely related to the injection of hot electrons into the gate oxide is the creation of substrate current. The substrate current is produced by hole flow (in NMOS) created from electron-hole pair generation by impact ionization in the drain-depletion region. The substrate current can cause source-substrate debiasing and hence device breakdown.

Even with these additional problems, the advantages of increased speed and circuit density still make scaling advantageous.

REFERENCES

1. V. L. Rideout, F. H. Gaensslen, and A. LeBlanc, "Device design consideration for implanted n-channel MOSFETs," *IBM J. Res. Dev.,* p. 50, January 1975.

2. J. R. Brews, "Threshold shifts due to nonuniform doping profiles in surface channel MOSFETs," *IEEE Trans. Electr. Dev.,* vol. ED-26, p. 1696, 1979.

3. S. M. Sze, *Physics of Semiconductor Devices,* Chap. 8, Wiley, New York, 1981.

4. M. B. Barron, "Low level currents in insulated gate field effect transistors," *Solid State Electron.,* vol. 15, p. 293, 1972.

5. W. Fichtner and H. W. Pitzl, "MOS modeling by analytical approximations. I. Subthreshold current and threshold voltage," *Int. J. Electron.,* vol. 46, p. 33, 1979.

6. L. A. Akers and J. J. Sanchez, "Threshold voltage models of short, narrow and small geometry MOSFETs: A review," *Solid State Electron.,* vol. 25, p. 621, 1982.

7. R. S. Muller and T. I. Kamins, *Device Electronics for Integrated Circuits,* Chap. 3, Wiley, New York, 1977.

8. R. C. Varshney, *Electr. Lett.,* vol. 9, no. 25, p. 600, 1973.

9. L. D. Yau, "A simple theory to predict the threshold voltage of short-channel IGFETs," *Solid-State Electr.,* vol. 17, pp. 1059–1063, 1974.

10. G. W. Taylor, "Subthreshold conduction in MOSFETs," *IEEE Trans. Electr. Dev.,* vol. ED-25, p. 337, 1978.

11. R. R. Troutman, "Subthreshold design considerations for IGFETs," *IEEE J. Solid State Circ.,* vol. SC-9, p. 55, 1974.

12. L. A. Akers, "Threshold voltage of a narrow-width MOSFET," *Electr. Lett.,* vol. 17, no. 1, pp. 49–51, 1981.

13. L. A. Akers, "An analytical expression for the threshold voltage of a small geometry MOSFET," *Solid-State Electr.,* vol. 24, pp. 621–627, 1981.

14. J. R. Brews, W. Fichtner, E. H. Nicollian, and S. M. Sze, "Generalized guide for MOSFET miniaturization," *IEEE Electr. Dev. Lett.,* vol. EDL-1, p. 2, 1980.

15. R. H. Dennard, F. H. Gaensslen, H. Yu, V. L. Rideout, E. Bassons, and A. R. LeBlanc, "Design of ion-implanted MOSFETs with very small physical dimensions," *IEEE J. Solid-State Circ.,* vol. SC-9, p. 256, 1974.

PROBLEMS

1. Derive the threshold voltage expression, Eq. (6-24), using Fig. 6-6.

2. Derive the threshold voltage expression, Eq. (6-27), using Fig. 6-7.

3. Calculate the threshold voltage for a device with $L = 1$ μm, $W = 2$ μm, $t_{ox} = 150$ Å, $N_A = 1.0 \times 10^{16}$ cm^{-3}, $r_j = 0.1$ μm, and the gate is aluminum.

4. For the data given in Prob. 3, except for $W = 100$ μm, calculate and plot the subthreshold current.

7

MOS IC Processes

7-1 METAL-GATE PMOS

The first MOS IC technology to go into commercial production was the metal-gate PMOS process. It utilized (111) wafers, also used by bipolar circuits of that time. This process has since been largely replaced by others but is discussed here to provide a historical perspective. With this process, MOS manufacturing started with very few mask layers and a relatively simple process. Other processes, to be discussed in later sections, have all seen increased process complexity and increased mask count.

The fabrication sequence will be explained with the aid of Fig. 7-1. First a thick layer of oxide is grown to serve as a mask for subsequent diffusion steps (step *a*). Then the first mask layer (source and drain) is used to etch the oxide over the source and drain regions (step *b*). Boron is then diffused in, forming the *p*-type source and drain junction (step *c*). Next a thick layer (13,500 Å) of oxide is either grown or deposited by chemical vapor deposition (step *d*). Holes are then etched by means of the second mask (gate and contact) for openings where the gate oxide will be and wherever there are contacts to be made to diffusion areas (step *e*). Gate oxide is then grown, covering all openings with a layer of quality oxide (step *f*). A third mask (contacts) is then used to reopen all the contacts to the diffusions (step *g*). Opening the contacts with two cuts allows metal lines to conform to the large step created by the contact holes with less likelihood of cracking. A two-step process also prevents the contact holes from "blowing" too large as the etchant works through the thick oxide. The final contact opening is determined only by the second mask and etch step which goes through the relatively thin gate oxide.

Metal, in the form of aluminum, is then evaporated over the whole wafer and later patterned with the fourth mask (metal). This layer simultaneously

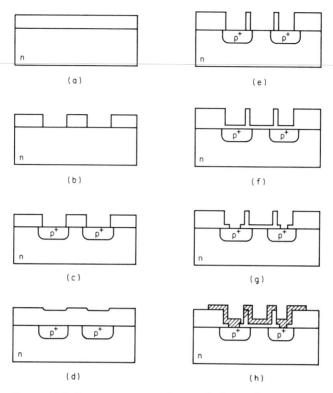

Figure 7-1 Process sequence for standard metal-gate PMOS process.

forms the gate electrode as well as the circuit interconnect pattern (step *h*). A metal anneal of about 450°C is performed at this point to anneal out x-ray damages during metal deposition and to form an alloy between the aluminum and the silicon it contacts to. Although not shown in Fig. 7-1, but needed in all process flows, a layer of phosphosilicate glass (PSG) is deposited as a passivation layer. It protects against mechanical scratches, moisture, and mobile alkali ions. A fifth mask (passivation) is then needed to access to metal bonding areas for assembly into final packages.

With a background doping concentration of $N_D = 10^{15}$ cm^{-3}, an oxide thickness of $t_{ox} = 1200$ Å, and a fixed interface charge density of $Q_f/q = 5 \times 10^{11}$ cm^{-2}, then $V_T = -4.1$ V (see Prob. 1). With a device threshold of -4 V, a supply voltage of three to four times that is needed, leading to the standard supply of -15 V for early MOS circuits.[1]

When a metal line crosses the source and drain regions of two different active devices, a parasitic MOS transistor structure is formed. Its threshold is defined as the field threshold and is -37 V for this process (see Prob. 1), roughly an order of magnitude higher than the active device threshold. This is expected since the field oxide that separates active devices is 13,500 Å, an order of magnitude thicker than the gate oxide. By designing the field threshold to be higher than the power supply voltage, the field region will not invert, even with the maximum operating voltage on the metal line. This maintains the isolation between devices and between diffusion lines.

7-2 HYPOTHETICAL METAL-GATE NMOS

Electron mobility is three times hole mobility; therefore for the same W and L, i.e., the same silicon area, the current driving capability of n-channel transistors is three times that of p channels. NMOSs thus have three times the speed of PMOS circuits. There is therefore a great impetus to employ NMOSs. Yet the PMOS process was the first to go into production because of problems with the NMOS process which will be discussed below.

Problem 2 shows that ϕ_{MS} and Q_f cause the threshold for a metal-gate NMOS process to be -0.34 V. A negative threshold for an n-channel device indicates depletion-mode operation, unsuitable for conventional logic circuits. This assumes that a switch to (100) material has already been made in anticipation of the need to reduce Q_f. A higher Q_f number would make matters even worse. Several ways of converting n-channel devices from depletion- to enhancement-mode are as follows:

1. *Substrate Bias* Back-biasing the $p\mu$ substrate -2 to -5 V with respect to the source, i.e., system ground, is enough to shift the threshold voltage to the enhancement-mode region. Early NMOS circuits did indeed use this technique, as illustrated in Fig. 7-2. The extra negative power supply is an obvious additional system cost that makes this undesirable as a long-term solution. However, a circuit technique has been invented that allows a substrate bias generator to be built on-chip (see "Substrate Bias Generator," in Sec. 9.6), replacing the external supply. This generator would supply only the leakage current of reverse-biased pn junctions.

2. *Increased Substrate Doping* By increasing the substrate doping, V_T will become positive. This is effectively what is done when n-channel devices in metal-gate CMOS are placed inside a p-tub. By its nature, a

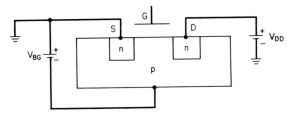

Figure 7-2 Providing substrate bias to NMOS circuits.

tub formed by diffusion is of higher doping than the background material. For straight NMOS circuits, this technique is not used because additional doping degrades mobility and increases drain capacitance.

3. *Silicon Gate*   If p-doped polysilicon is used as the gate electrode instead of aluminum, ϕ_{MS} changes from -0.9 V to $+0.25$ V according to Fig. 3-10. This moves the threshold from -0.34 V to $+0.81$ V, an enhancement value. Unfortunately, the use of p-type polysilicon is not compatible with the subsequent n-type source and drain doping. Neither does it yield the lowest sheet resistance for polysilicon. Hence this technique has not gained acceptance.

4. *Threshold Adjust*   Boron can be ion-implanted to adjust the threshold by about 1 V to change it to an enhancement value and, for that matter, to any nominal value in the enhancement range. With the advent of ion implantation, this technique is by far the most practical and most commonly used.

7-3   METAL-GATE CMOS

To provide both p and n channels for CMOS circuits requires both n- and p-type silicon as background. This incompatibility is solved by forming a well or tub of one type into the other. The most common procedure is the use of p-tub into n substrate, but the use of n-well into p-type substrate is also feasible and is covered in Chap. 11. Figure 7-3 depicts the process sequence[2] for a p-tub metal-gate CMOS. A thick oxide is grown and the first mask (p-well) opens the p-tub regions (step a). The p-type dopant is introduced by depositing boron in a diffusion predeposition cycle or by ion implantation. Boron is then driven in by a diffusion cycle that includes oxidation (step b). Because of the faster oxidation rate where bare silicon is exposed, more silicon is consumed in the p-tub region. This generates a step in the silicon that subsequent masking layers can align to. The p-tub is driven into a depth of about 5 μm, which is calculated to be deep enough to avoid punch-through between the substrate and the n-channel drains.

The second-layer mask (p^+ source and drain) cuts openings for boron diffusion (step c). The boron diffusions outside the p-tub form p-channel source and drain or p^+ cross-under diffusions, while those inside form contacts to the tub (step d). The third layer (n^+ source and drain) similarly cuts holes for the phosphorus diffusion (step e). This time, contacts to the n-substrate, n-channel source and drains and n^+ cross-unders inside the tub are formed (step f). The fourth layer (gate and contacts) opens the gate and contact regions (step g). After the gate oxide is grown, the fifth layer (contact) reopens all the contact areas (step h). Metal is deposited and patterned with the sixth layer (metal) forming the gate electrodes and circuit interconnects (step i). Finally, passivation glass is deposited and holes to bonding pads are opened with a seventh layer (passivation).

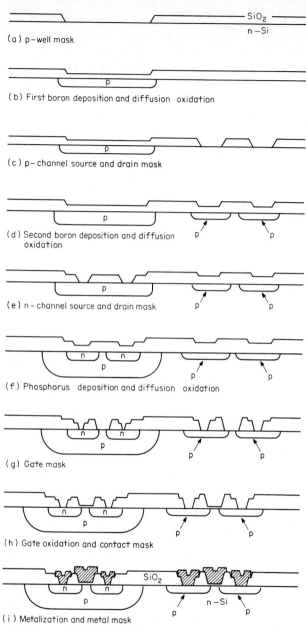

(a) p-well mask

(b) First boron deposition and diffusion oxidation

(c) p-channel source and drain mask

(d) Second boron deposition and diffusion oxidation

(e) n-channel source and drain mask

(f) Phosphorus deposition and diffusion oxidation

(g) Gate mask

(h) Gate oxidation and contact mask

(i) Metalization and metal mask

Figure 7-3 □ Metal-Gate CMOS process. (*a*) *p*-Well mask; (*b*) first boron deposition and diffusion oxidation; (*c*) *p*-channel source and drain mask; (*d*) second boron deposition and diffusion oxidation; (*e*) *n*-channel source and drain mask; (*f*) phosphorus deposition and diffusion oxidation; (*g*) gate mask; (*h*) gate oxidation and contact mask; (*i*) metalization and metal mask. (*After Ref. 2*)

The metal-gate CMOS process is also a relatively simple one, requiring only seven masking steps. Process simplicity translates into lower cost per finished wafer. However, the process does have the usual disadvantages associated with metal gate, such as poor layout packing density and low speed. The next few sections will show that more elaborate processes can overcome these disadvantages. With regard to layout density, the metal-gate CMOS process suffers yet another shortcoming. To avoid leakage paths in the field regions and to lower the probability of latch-up (see Sec. 11-6), each device is surrounded by a stripe of "guard ring" diffusion. n-Channel devices are surrounded by p^+ and p-channel devices by n^+. This procedure increases the area of a given circuit by a significant factor.

7-4   SILICON-GATE LOCOS NMOS PROCESS

In a metal-gate process, the gate oxide region is drawn to overlap the source and drain, as in Fig. 7-4, assuring that even with misalignment between masking operations, a continuous channel is present from source to drain. The metal electrode, in turn, is also drawn with overlap to the gate region so as not to leave gate regions uncovered by metal. All the allowances for misalignment enlarge the area needed to lay out the circuit. In addition, the thin gate oxide in the overlap region produces a large overlap capacitance between the gate and the drain, slowing a circuit considerably. Its equivalent capacitive loading to the previous logic stage is the overlap capacitance multiplied by a factor of 1 plus the gain of the transistor (Miller effect). Both problems can be solved with a self-aligned process where the gate serves as a mask for the source and drain diffusions. The gate electrode has to be of refractory material, i.e., one that can withstand the subsequent high-temperature diffusion steps. Polysilicon fits the requirement and is universally used for self-aligned processes.[3]

Polysilicon is short for polycrystalline silicon, a deposited layer of silicon with grains of single-crystal silicon a few microns wide, separated by thin grain boundaries. Being silicon, it can also withstand high processing tem-

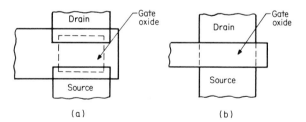

Figure 7-4 Alignment tolerances on metal-gate processes result in excess circuit area when compared to the self-aligned silicon gate process. (*a*) Metal gate; (*b*) self-aligned polysilicon gate.

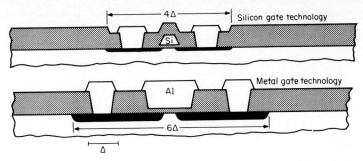

Figure 7-5 Cross-sectional areas of (*a*) self-aligned gate vs. (*b*) metal-gate structures.

peratures and can be oxidized or doped with impurities. Polysilicon is chosen over certain refractory metals such as tungsten or titanium because a thinner layer (about 4500 Å) can be deposited, making it easier to pattern into finer lines.

The net improvement in overall area from the use of a self-aligned process is dramatic, as Fig. 7-5 shows.

The silicon-gate NMOS process is illustrated in Fig. 7-6. The process sequence includes a widely used step called LOCOS (localized oxidation of silicon[4]) that produces a fairly planar surface. A surface that is free of high, sharp steps is conducive to resolving and patterning very dense features on an integrated circuit.

First, a thin layer (700 Å) of oxide is grown, followed by the deposition of a silicon nitride (Si_3N_4) layer of comparable thickness (1000 Å). The first masking layer (active area) is used to pattern the nitride leaving nitride over active areas. The active areas include the diffusion areas (all n^+) and the gate oxide or channel regions. After the nitride is patterned, boron is ion-implanted into the field region outside the active area, while prevented from penetrating the nitride (step *a*). This implant raises the surface doping concentration in the field region and subsequently acts to raise the field threshold voltage. At this point the wafer is subjected to an oxidation cycle to grow 9000 Å of field oxide (step *b*). A thin top layer of the nitride is oxidized, but since nitride itself is highly impervious to oxygen, the silicon underneath is not affected by the oxidation. At the edges of the nitrite layer, lateral diffusion of oxygen produces a tapered oxide which resembles a bird's beak when examined at high magnification (see Fig. 7-7). The fact that half of the field oxide thickness lies underneath the original surface and that the edges are tapered results in a planar surface. Note that there is a slight, about 1 μm per side, "encroachment" of the field oxide into the active area. The ratio of the nitride layer thickness to the original oxide thickness strongly influences the amount of encroachment and the residual stress at the edges.

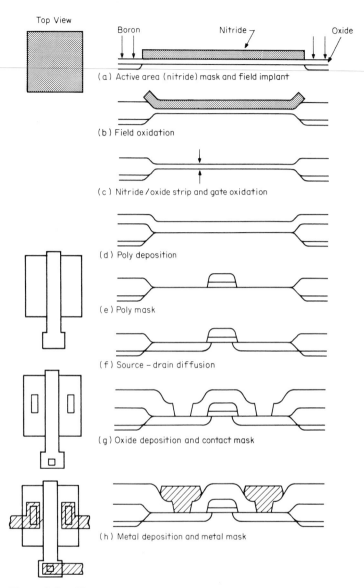

Figure 7-6 Silicon-gate LOCOS NMOS process. (*a*) Active area (nitride) mask and field implant; (*b*) field oxidation; (*c*) nitride-oxide strip and gate oxidation; (*d*) poly deposition; (*e*) poly mask; (*f*) source and drain diffusion; (*g*) oxide deposition and contact mask; (*h*) metal deposition and metal mask.

The nitride and underlying oxide is stripped, and 1000 Å of quality gate oxide is grown (step *c*). At this point, a light boron implant, for example, 5 × 10^{11} cm^{-2} at 30 keV, is used to set the threshold voltage of enhancement-mode devices. The second mask (depletion) is used to open the areas where a shallow phosphorus or arsenic implant can create depletion devices. Polysilicon is deposited (step *d*) and patterned by the third mask (poly) to form gate electrodes and a separate layer of interconnect (step *e*). Polysilicon is a poor, albeit satisfactory, conductor; the availability of an additional layer of interconnect undoubtedly expedited the acceptance of the silicon gate process.

The source and drain diffusions (or implantations) are then performed (step *f*). Next, a layer of phosphorus-doped glass is deposited to form the

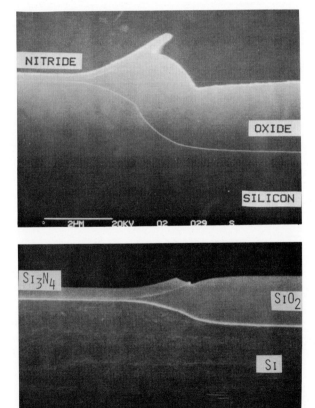

Figure 7-7 Encroachment of field oxide into active area produces bird's beak effect: (*a*) with silicon trench before oxidation, (*b*) without. (*Photos courtesy of Simon Thomas, Burroughs Corp.*)

interlayer dielectric, insulating the polysilicon from the metal which is to be placed over it. But before the metal is deposited, contact holes are etched with the fourth mask (contacts) (step *g*). Contacts can be made between metal and diffusion as well as between metal and polysilicon. A step inserted at this point is to subject the opened contacts to contact enhancement. A phosphorus diffusion or implantation forms a better ohmic contact between the metal and the silicon; but it also forms a deeper junction as Fig. 7-8

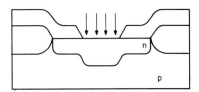

Figure 7-8 **Deeper junction formed by contact enhancement.**

shows. This lessens the likelihood of metal, in alloying with the silicon, from "spiking" through and shorting out the junction.

After all residual oxide is removed from the contact holes, metal is deposited and patterned by the fifth mask (metal) (step *h*). As usual, this is followed by metal anneal and alloy, passivation deposition, and the last mask (passivation) for bonding pads. It is seen that the number of masking steps can be small even when switching to the silicon-gate process; however, most silicon-gate NMOS processes have added embellishments such as buried contacts and oversized contacts (see Sec. 7-6) such that the final mask layer count is invariably higher than for a metal-gate process.

7-5 HMOS PROCESS

The HMOS (high-performance MOS) process[5,6] is built upon silicon-gate NMOS and simultaneously increases speed and density by applying the principle of device scaling. The gate length is simply reduced from 5 to 3.5 μm, but there are other key changes in the process that make this reduction in gate length feasible. The process flow is very similar to the silicon-gate NMOS LOCOS process; therefore only the differences will be discussed below.

The first change in the process flow is the use of a high-resistivity *p*-type substrate, on the order of 50 $\Omega \cdot$cm. This lowers the junction capacitance, reduces the ΔV_T due to back-gate (substrate) bias, and increases the surface carrier mobility. All three factors result in faster, lower power circuits. Next, the gate oxide is reduced from 1000 to 700 Å. This increases the device gain β, while avoiding the deleterious short-channel effects such as drain-to-source punch-through. At the same time, the source and drain junction depth is reduced to 0.8 μm from 2.0 μm by switching from phosphorus as a dopant to the slower-diffusing arsenic. Arsenic doping can be accomplished by spinning on a viscous solution of arsenic-doped glass (very much akin to applying photoresist) or by ion implantation. In both cases a long 2- to 3-h drive-in at 900°C is necessary to lower the sheet resistance to the desired range. This long drive-in results in the 0.8-μm junction depth. The shallow junction reduces periphery junction capacitance and the gate-to-drain overlap capacitance, while improving density by allowing smaller diffusion-to-diffusion spacing.

The HMOS process relies extensively on ion implantation to set threshold voltages. In fact there are four different device thresholds to choose from for added design flexibility. Figure 7-9 shows how the four thresholds are obtained by the binary combination of the presence or absence of two implants. Without any implants, a device has a natural threshold of around 0 V. With a boron implant, the device is shifted to an enhancement threshold of 0.8 V. With a phosphorus implant alone, there is a shift to a depletion threshold of −2.4 V. With both implants, there is a net shift to −1.5 V, still a depletion device. The enhancement devices are used for normal logic circuitry. The −2.4-V depletion devices are used as load devices which need a larger current drive capability. The −1.5-V devices are used as load devices where a lower power dissipation is desired. The 0-V threshold devices are used for transmission devices which should have low threshold drops, but the threshold should not be negative so that the device can still be shut off.

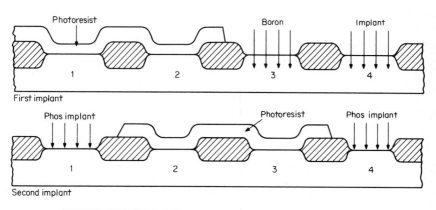

	Device results		
Device no.	Boron implant	Phos implant	V_{TO} @ $V_{BG} = -2.5$, V
1	No	Yes	−2.4
2	No	No	0
3	Yes	No	+0.8
4	Yes	Yes	−1.5

Figure 7-9 Four threshold voltages are obtained from two implants for the HMOS process. Top, first implant; bottom, second implant. (*Motorola, Inc.*)

7-6 PROCESS ENHANCEMENTS

Modern MOS processes contain many process steps that add to the complexity of the overall sequence, but enhance the density, yield, and/or performance of the circuit. An example was already given in the case of HMOS, where two additional photomasking and threshold implant steps result in added design flexibility. Other examples will be discussed below.

Buried Contacts

In NMOS and HMOS, the polysilicon gate of depletion load devices is tied to its source. The gate oxide normally insulates the gate from the source and the drain, but by adding the buried contact masking step, openings are etched in the gate oxide before the polysilicon is deposited to provide a natural short between gate and source. Significant savings in area can be obtained.

The process sequence is shown in Fig. 7-10. Step *a* shows an opening in the gate oxide from the buried contact mask. The polysilicon is deposited and doped in step *b*. The doping process causes a junction to be diffused into the silicon. In subsequent processing, step *c*, the usual source and drain diffusion is formed. The polysilicon forms a good ohmic contact with the source and drain diffusion. Note that the junction formed by doping from the polysilicon ends up being deeper (and diffuses further laterally) than the source and drain junctions. Layout of circuits using buried contacts needs to take this fact into account.

The deeper junction formed by the buried contact is put to good use in forming a low-resistance diffusion cross-under. If a buried contact is opened throughout the length of a diffusion run and topped with a similar length of polysilicon run, there is a 30 to 50 percent reduction in series resistance when compared to using either diffusion or polysilicon alone.

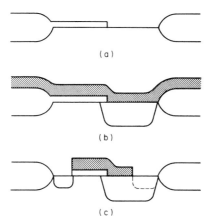

(a)

(b)

(c)

Figure 7-10 Process sequence for buried contacts.

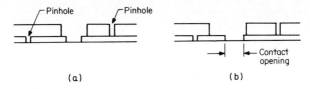

Figure 7-11 Oversized contacts allow for misalignment. (*a*) Perfect alignment; (*b*) with misalignment.

Oversized Mask

To pattern very fine features such as contacts or buried contacts, a thinner layer of photoresist is called for. The reduced thickness makes the photoresist more susceptible to pinholes and breaks in going over steps on the wafer. Etchants would go through these defects just as readily as through normal contact openings, causing short circuits. One solution is to spin on a second layer of photoresist after the first layer has been exposed and developed, but before etching takes place. The second layer is exposed with a mask that has the same pattern as before, though with all the contacts enlarged (oversized). The idea is that random pinholes in the layers seldom line up (see Fig. 7-11). The second layer is oversized, because even with misalignment between the two, the first opening is still fully uncovered as the figure shows. Note that the first layer is the one that determines the size of the etched opening.

Metal Silicides

As MOS integrated circuits advance in speed by shrinking to smaller geometries, their performance is increasingly limited by the RC time constant delay of polysilicon interconnects. Their sheet resistance is typically 15 to 30 Ω per square and will go no lower than 10 Ω per square even with degenerate doping. An attractive solution is to form a top layer of metal silicide on the polysilicon to bring the effective sheet resistance down to the 1 or 2 Ω per square range. This solution is preferred over that of replacing polysilicon directly with metal because the basic polysilicon-gate MOS structure is preserved.

The silicide[7] is formed by first depositing a thin layer of refractory metal such as platinum, titanium, tantalum, molybdenum, or tungsten. When heated to a sintering temperature of around 900°C, a layer of silicide is formed. Unreacted metal that is contacting oxide rather than silicon can be etched away, e.g., platinum in hot aqua regia. What is left behind is the original layer of polysilicon with its top surface converted to silicide, e.g., titanium disilicide, $TiSi_2$, or platinum silicide, $PtSi$.

This procedure has been used extensively since the mid-1960s in bipolar devices to form Schottky barrier diodes for Schottky-TTL circuits; but recent efforts are directed toward forming the silicide on polysilicon rather than on single-crystal silicon. Research is continuing on alternative techniques for silicide formation[8] (such as cosputtering) to control the metal-to-silicon ratio in the silicide, on the ease in patterning the silicide-polysilicon sandwich, and on the stability and compatibility of the silicide layer with subsequent IC processing.

Other discussions of the processes described in this chapter, as well as other MOS processes, are available in Refs. 9 and 10.

REFERENCES

1. W. N. Carr and J. P. Mize, *MOS/LSI Design and Application,* Chap. 2, McGraw-Hill, New York, 1972.

2. W. M. Penney (ed.), *MOS Integrated Circuits,* Chap. 3, Van Nostrand Reinhold, New York, 1972.

3. F. Faggin and T. Klein, "Silicon gate technology," *Solid-State Electr.,* vol. 13, pp. 1125–1144, 1970.

4. J. A. Appels et al., "Local oxidation of silicon and its application in semiconductor technology," *Philips Research Reports,* vol. 25, pp. 118–131, April 1970.

5. R. Pashley et al., "H-MOS scales traditional devices to higher performance level," *Electronics,* Aug. 18, 1977.

6. R. M. Jecmen, "HMOS II static RAMs overtake bipolar competition," *Electronics,* Sept. 13, 1979.

7. J. G. Posa, "Silicides nudging out polysilicon," *Electronics,* pp. 101–102, Nov. 3, 1981.

8. S. P. Murarka, "Refractory silicides for integrated circuits," *J. Vac. Sci. Technol.,* vol. 17, no. 4, pp. 775–792, July–August 1980.

9. A. B. Glaser and G. E. Subak-Sharpe, *Integrated Circuit Engineering,* Chap. 6, Addison-Wesley, Reading, Mass., 1977.

10. D. J. Hamilton and W. G. Howard, *Basic Integrated Circuit Engineering,* Chap. 6, McGraw-Hill, New York, 1975.

PROBLEMS

1. For a metal-gate PMOS process, calculate the threshold voltage for

   $$N_D = 10^{15} \text{ cm}^{-3}$$

   $$t_{ox} = 1200 \text{ Å}$$

   $$T = 300 \text{ K}$$

   $$Q_f/q = 5 \times 10^{11} \text{ cm}^{-2}$$

   Repeat the calculations for the field device whose field oxide $t_{ox} = 13,500$ Å.

2. For the hypothetical metal-gate NMOS process, calculate the threshold voltage for

   $N_A = 10^{15} \text{ cm}^{-3}$

   $t_{ox} = 1200 \text{ Å}$

   $T = 300 \text{ K}$

   $Q_f/q = 9 \times 10^{10} \text{ cm}^{-2}$

   Anticipating a need to lower Q_f, the switch to (100) material is assumed to have been made.

3. What would happen to aluminum if it were used as self-aligned gate material and were subjected to 900°C source and drain diffusion? Have specific data to back up your answer.

4. The contact enhancement for the NMOS process allows contacts to be positioned closer to the edge of a diffusion. Imagine a contact that is actually intersecting the edge of a diffusion due to misalignment. Would the contact short out the diffusion to the substrate? Explain your answer.

8

MOS
Wafer Fabrication

8-1 OXIDATION

Oxidation is the process by which a layer of silicon dioxide (SiO_2) is formed on the surface of silicon. The oxidizing species diffuses through the already grown oxide to react with the silicon at the $Si-SiO_2$ interface. For a given oxide thickness t_{ox} grown, the oxide surface will be at $0.54t_{ox}$ above the original silicon surface, while the oxide-silicon interface will be $0.46t_{ox}$ below[1] (see Fig. 8-1).

The general equation for oxidation is given by[2]

(8-1) $\quad t_{ox}^2 + At_{ox} = B(t + \tau)$

or

(8-2) $\quad t_{ox} = \dfrac{A}{2}\left(\sqrt{1 + \dfrac{t + \tau}{A^2/4B}} - 1 \right)$

where t_{ox} = oxide thickness
$\quad\quad B$ = parabolic rate constant
$\quad B/A$ = linear rate constant
$\quad\quad t$ = oxidation time
$\quad\quad \tau$ = initialization parameter

The generalized nature of Eq. (8-1) is clearly seen in Fig. 8-2 where the results of various oxidation conditions are plotted on the normalized axis of

$$\frac{t + \tau}{A^2/4B} \quad \text{versus} \quad \frac{t_{ox}}{A/2}$$

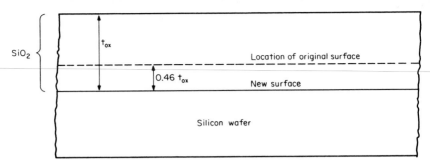

Figure 8-1 New silicon surface after oxidation. (*After Ref. 1. Used with permission of McGraw-Hill Book Company.*)

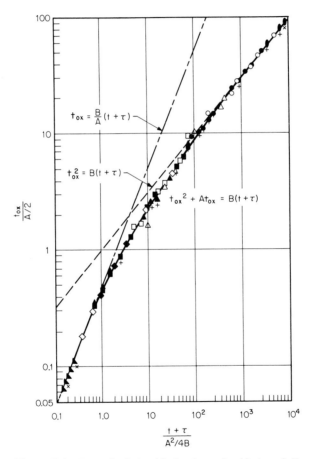

Figure 8-2 General relationship for thermal oxidation of silicon. ●, 1300°C; ○, ●, 1200°C; △ ▲, 1100°C; □, ■, 1000°C; ◇, ◆, 920°C, ▲, 800°C; □. Filled shapes refer to O_2; empty shapes refer to H_2O. (*After Ref. 2.*)

150

There are two regions in the plot. With thin oxides and at low temperatures, the growth rate is limited by the surface reaction rate at the $Si-SO_2$ interface and the oxide thickness is a linear function of the oxidation time:

$$(8\text{-}3) \quad t_{ox} = \frac{B}{A}(t + \tau)$$

At higher temperatures and for thicker oxides, the reaction is limited by diffusion of the oxidizing species through the previously formed oxide. In this case, the oxide thickness is proportional to the square root of the oxidation time:

$$(8\text{-}4) \quad t_{ox} = \sqrt{B(t + \tau)} \cong \sqrt{Bt}$$

Oxides can be grown either with dry or wet oxidation. Dry oxidation uses O_2 gas as the oxidizing ambient. The resulting oxides are of better dielectric quality and have lower defect density, hence critical steps such as gate oxidation almost invariably use dry oxidation. A graph of oxidation thickness grown in dry oxidation is shown in Fig. 8-3 as a function of oxidation time for various oxidation temperatures.[3] Wet oxidation is performed by bubbling oxygen through deionized water kept near the boiling point at 97°C or by forming a "torch" (flame) with hydrogen and oxygen gases directly in the oxidation furnace. Wet oxidation allows a much faster oxide

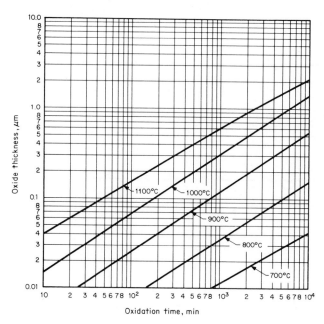

Figure 8-3 Oxide growth in dry oxygen for (100) silicon. (*After Ref. 3.*)

growth rate since H_2O diffuses through the oxide at a much faster rate than pure oxygen alone. This allows a thicker layer of oxide to be grown in a shorter length of time. Conversely, the same layer can be grown for longer time periods but at a low enough temperature such that there is minimal redistribution of previously introduced dopants. Because of these reasons, wet oxidation is often used in the growth of the thick field oxide between active devices. A graph of oxide thickness grown in steam is shown in Fig. 8-4.

When a wafer with a layer of oxide already present is reoxidized, the initial thickness is translated into an equivalent initial oxidation time for the subsequent oxidation. The additional oxidation time is added to this initial value for a total oxidation time. An example is given below.

Example: A wafer undergoes a steam oxidation cycle for 40 min at 900°C and then is subjected to a dry oxidation cycle for 700 min at 1000°C. Find the final oxide thickness.

Solution: From Fig. 8-4 for wet oxidation, we find that 40 min of wet oxidation at 900°C produces 1000 Å (0.1 μm) of oxide. Referring to Fig. 8-3 for dry oxidation, we find that 1000 Å of oxide is equivalent to an initial oxidation time of 200 min at 1000°C. An additional time of 700 min gives a total time of 900 min. At 1000°C, this yields a final thickness of 3000 Å.

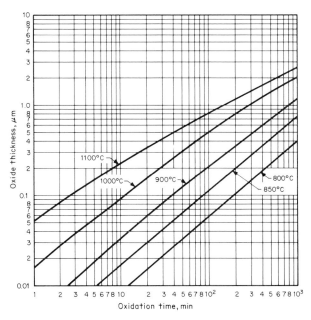

Figure 8-4 **Oxide growth in pyrogenic steam for (100) silicon.** (*After Ref. 3.*)

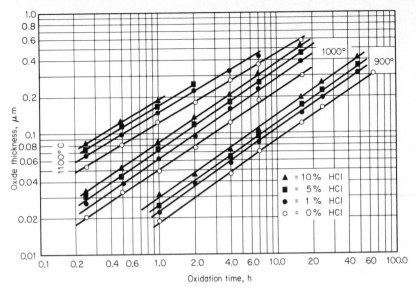

Figure 8-5 Effect of adding HCl during oxidation on (100) Si growth rate. (*After Ref. 3.*)

The introduction of a few percent (1 to 9% by volume) of HCl to the oxygen during oxidation improves the film quality, acts as a gettering agent for sodium in the oxide, and improves a variety of other device parameters.[4] For this reason, it is common practice to incorporate HCl or another chlorine-bearing species, such as trichloroethane,[5] during the gate oxidation. Incorporation of HCl during oxidation does increase the oxidation rate by approximately 30 percent. This is evident from Fig. 8-5, which gives the result of varying the amount of HCl in the oxidation process.

Redistribution of Impurities during Oxidation

When extrinsic silicon undergoes an oxidation cycle, the impurity concentration near the surface redistributes. The amount and manner of redistribution depends on the segregation coefficient, defined as

$$m = \frac{\text{equilibrium concentration of impurity in silicon}}{\text{equilibrium concentration of impurity in } SiO_2}$$

It was shown in Fig. 4-22 that for slow-diffusing impurities such as boron where $m < 1$, there is a depletion of impurities as the growing oxide *takes up* the impurity. For other slow-diffusing impurities such as phosphorus where $m > 1$, there is a pileup of impurities as the growing oxide *rejects* the impurity. Note that even if $m = 1$, some redistribution of impurity will still take place. SiO_2 takes up more volume than the silicon used in the oxidation; thus the same amount of impurity is distributed in a larger volume, and the concentration near the surface is lower.

Figure 8-6 shows the amount of redistribution in boron concentration for various oxidation temperatures.[6] Figure 8-7 shows the corresponding graph for phosphorus. The disparity between the concentrations at the surface and in the bulk is greatest for low temperatures and for wet oxidation. Figures 8-6 and 8-7 represent the steady-state condition and are independent of oxidation time and oxide thickness.

High-Pressure Oxidation

If oxidation is carried out under high pressure (greater than 1 atm), the growth rate increases in direct proportion to the pressure.[7] This can be put to great use in growing thick oxides at low temperatures to avoid significant redistribution of previous diffusions. Oxidation at lower temperature also results in denser oxides with higher indexes of refraction and lower etch rates. High-pressure oxidation also reduces the lengths of oxidation-induced stacking faults (OSF) as well as their density.[8] OSFs are defects in the silicon that are generated during oxidation when excess interstitial silicon atoms precipitate to form extra atomic planes. These defects are visible when the oxide is stripped and the silicon surface is subjected to certain preferential (anisotropic) etchants. The OSFs, "decorated" for easy identification, appear as dumbbell-shaped structures tens of microns long. OSFs cause excessive junction leakage, excessive noise, and other device degrada-

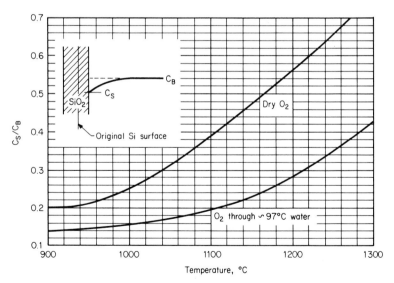

Figure 8-6 Redistribution of boron during thermal oxidation. $C_B \cong 10^{16}$ cm^{-3}. *(After Ref. 6.)*

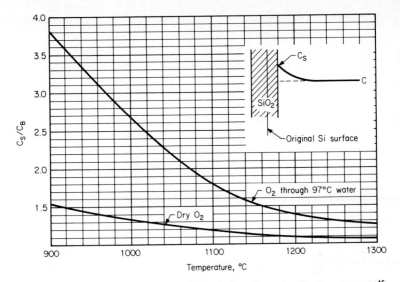

Figure 8-7 Redistribution of phosphorus during thermal oxidation. $C_B \cong 10^{16}$ cm^{-3}. *(After Ref. 6.)*

tion. The use of high-pressure oxidation has demonstrated significant leakage current reduction as evidenced by a factor of two in improvement in the time between refresh cycles of a MOS dynamic random access memory.[9]

High-pressure oxidation is found to be very much applicable to the LOCOS process.[9] The bird's beak encroachment decreases with increased temperature and increased ratio of nitride to bottom oxide thicknesses but is not affected by pressure. More nitride is oxidized under high pressure which must be taken into account, but otherwise, other device properties such as oxide-semiconductor interface charges behave as in the atmospheric pressure case.

High-pressure oxidation is carried out in a normal oxidation quartz tube, but with the whole furnace enclosed in a long steel pressure vessel. Although research studies have been conducted at up to 500 atm, for extra safety margin, commercial production is limited to 6 to 25 atm.

8-2 DIFFUSION[1]

To describe the diffusion of a collection of particles in a medium, consider the following definitions:

N = particle density
D = diffusion constant
f = particle flux density

Fick's first law states that the particle flux density is related to the gradient of the particle density by

(8-5) $\quad f = -D \, \nabla N$

For a one-dimensional case

(8-6) $\quad f = -D \dfrac{dN}{dx}$

Fick's second law states that the time rate of change of the particle density is in turn related to the divergence of the particle flux density:

(8-7) $\quad \dfrac{dN}{dt} = -\nabla \cdot f$

For a one-dimensional case

(8-8) $\quad \dfrac{dN}{dt} = -\dfrac{df}{dx}$

Combining Eqs. (8-6) and (8-8) gives

(8-9) $\quad \dfrac{dN}{dt} = -\dfrac{d}{dx}\left(-D \dfrac{dN}{dx}\right) = D \dfrac{d^2N}{dx^2}$

To solve Eq. (8-9), the boundary conditions for two specific cases will be applied.

In *predeposition*, dopant atoms are deposited on the surface of the wafer. The dopant source can be either gaseous, liquid, or solid. Though the source can be considered infinite, there is a maximum concentration of dopant atoms which can dissolve in silicon. That maximum is the solid solubility limit N_o, which differs for different impurities, but is relatively constant over a wide range of temperature. Sample values of N_o are 5.0×10^{20} cm^{-3} for boron, 1.0×10^{21} cm^{-3} for phosphorus, and 1.5×10^{21} cm^{-3} for arsenic, valid from 1050 to 1250°C. For the predeposition case, the boundary conditions for $t = 0$ and $x = 0$ are, respectively,

(8-10) $N(x, 0) = 0 \qquad N(\alpha, t) < \alpha$

(8-11) $N(0, t) = N_o u(t)$

where $u(t)$ is the unit step function. Figure 8-8 illustrates the boundary condition for this case. The solution, after applying the boundary conditions, is

(8-12) $\quad N(x, t) = N_o \, \mathrm{erfc}\left(\dfrac{x}{2\sqrt{Dt}}\right)$

Complementary error function is defined as

$$(8\text{-}13) \quad \text{erfc}\,(y) \equiv 1 - \text{erf}\,(y) = 1 - \frac{2}{\sqrt{\pi}} \int_0^y e^{-\lambda^2}\, d\lambda$$

Thus Eq. (8-12) can be written out fully as

$$(8\text{-}14) \quad N(x,\, t) = N_o\left(1 - \frac{2}{\sqrt{\pi}} \int_0^{x/2\sqrt{Dt}} e^{-\lambda^2}\, d\lambda\right)$$

Figure 8-9 gives a normalized plot of Eq. (8-12).

It is useful to calculate the total number of dopant atoms per unit area after a predeposition time t_p:

$$(8\text{-}15) \quad Q(t = t_p) = \int_0^\infty N(x,\, t_p)\, dx$$

$$= N_o \int_0^\infty \text{erfc}\left(\frac{x}{2\sqrt{Dt_p}}\right) dx$$

Tables of integrals show that

$$\int_0^\infty \text{erfc}\,(y)\, dy = \frac{1}{\sqrt{\pi}}$$

Therefore

$$(8\text{-}16) \quad Q(t_p) = \frac{2N_o}{\sqrt{\pi}}\,\sqrt{Dt_p}$$

After a thin layer of impurities is deposited by predeposition, the impurity source is removed and the wafer heated to a higher temperature to drive the impurities deeper. This is the *drive-in* diffusion. A drive-in cycle is called for whenever the surface concentration is required to be much less than N_o or when desired junctions are deeper than can be provided by a predeposition cycle. A good example is the p-tub in a CMOS circuit.

Oxygen can be and often is introduced during drive-in, because an oxide layer is usually required for subsequent photoresist steps.

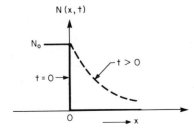

Figure 8-8 Boundary conditions for the predeposition diffusion.

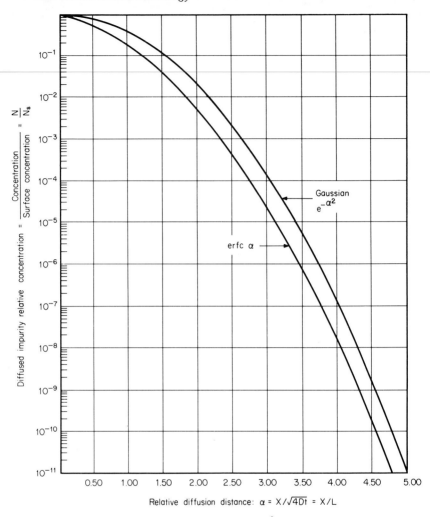

Figure 8-9 Normalized curves for erfc (α) and $e^{-\alpha^2}$, corresponding to predeposition and drive-in diffusions.

The boundary conditions for the drive-in cycle are

1. $N_A(\infty, t) = 0$, i.e., the wafer is assumed to be semi-infinitely thick.

2. No flux crosses the surface during drive-in, i.e., all of the predeposited impurities stay in the wafer. Mathematically,

(8-17) $\quad -D\dfrac{dN(0, t)}{dx} = 0$

3. The third item is more of a simplifying assumption than a boundary condition, that is, the predeposited atoms form a delta function at the surface:

(8-18) $\quad N(x, 0) = Q(t_p)\, \delta(x)$

The solution for the drive-in case is in the form of a *gaussian function:*

(8-19) $\quad N(x, t) = \dfrac{Q(t_p)}{\sqrt{\pi\, Dt}} \exp\left(\dfrac{-x^2}{4Dt}\right)$

A plot of Eq. (8-19) in normalized form is shown also in Fig. 8-9. Both curves have the same normalizing factor $2\sqrt{Dt}$ for x, but it must be understood that D and t are different sets of values for the two cases. If the dopants for a drive-in cycle come from a predeposition, then $Q(t_p)$ for Eq. (8-19) is given by Eq. (8-16).

Problem 3 goes through a typical predeposition-then-drive-in sequence and shows that the final junction depth is often an order of magnitude larger than the predeposition junction depth, justifying the initial assumption of a delta function for the predeposition charge.

The diffusion coefficient D is in units of square micrometers per hour (or square micrometers per second). It is a very strong function of temperature as Fig. 8-10 shows. The relationship is of the form

(8-20) $\quad D = D_o \exp \dfrac{-E_a}{kT}$

where D_o is a proportionality constant and E_a is the activation energy. If the wafer is subjected to several heat cycles, each with a corresponding D_i and t_i, then the equivalent Dt is simply the sum of all $D_i t_i$'s. This comes from the fact that solutions of the diffusion equations allow the use of an average diffusion coefficient D defined such that

(8-21) $\quad \overline{Dt} = \left(\dfrac{\sum\limits_i D_i t_i}{\sum\limits_i t_i}\right) \sum\limits_i t_i = \sum\limits_i D_i t_i$

In-process monitoring calls for test wafers to be included with the main lot while going through a diffusion step. The diffusion can then be evaluated by using the test wafers, often with measurements that are destructive. The two most commonly performed tests are for junction depth and sheet resistance.

Junction depth can be measured by lapping a bevel of about $1°$ to the wafer (see Fig. 8-11a). A stain such as concentrated HF is applied in the presence of intense light, causing p^+ regions to be darker than n^+ regions. When a monochromatic light is shone through a partially reflecting mirror,

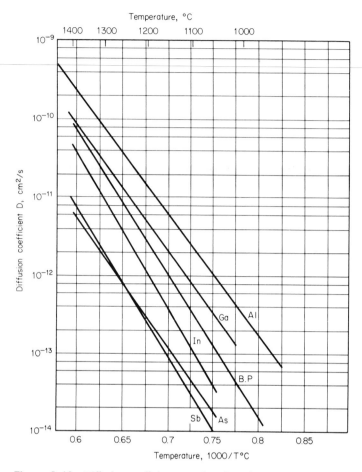

Figure 8-10 Diffusion coefficient as a function of temperature. (*After Ref. 19.*)

interference patterns are produced. The number of fringes from the wafer surface to the junction, when multiplied by half the wavelength, is the junction depth. It should be noted that the p^+ is stained only to the edge of the depletion layer and not to the metallurgical junction. This technique turns out to be independent of the actual bevel angle, a fact that led to the groove and stain technique, where instead of the painstaking beveling process, a groove is cut into the wafer with a special abrasive wheel. The resulting structure shown in Fig. 8-11*b* allows the same type of information to be read from the interference patterns.

The next most important monitor of a diffusion process is the measurement of sheet resistance. The four-point probe technique is most commonly used. Four close but equally spaced probes are brought down on the monitor wafer. A direct current is passed through the two outer probes as shown in Fig. 8-12, while the voltage drop across the inner probes is measured. If the wafer diameter is large in relation to the probe spacing (which in turn is larger than the layer thickness), then the sheet resistance is given by

$$(8\text{-}22) \quad \rho_s = \frac{\pi}{\ln 2} \frac{V}{I} = 4.53 \frac{V}{I}$$

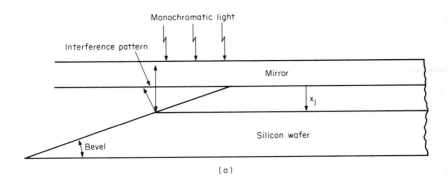

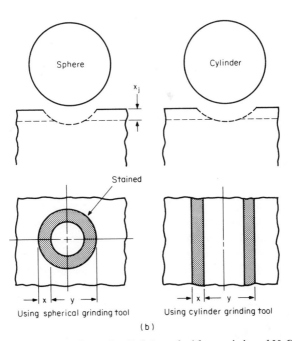

Figure 8-11 (a) Beveling (*After Ref. 1, used with permission of McGraw-Hill Book Company.*) and (b) grooving for junction-depth measurement, (*After W. R. Runyan, Semiconductor Measurements and Instrumentations, McGraw-Hill, New York, 1975.*)

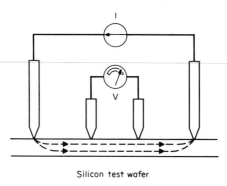

Figure 8-12 Measuring sheet resistance with the four-point probe technique. *(After Ref. 1. Used with permission of McGraw-Hill Book Company.)*

Silicon test wafer

The four-point probe technique overcomes the problem of contact series resistance that could affect a two-point voltage-current measurement.

A sheet resistance measurement pattern can also be included on an integrated circuit to provide measurement on a completed chip rather than a test wafer. The diffusion pattern can be as simple as a two-terminal rectangular strip as shown on the left half of Fig. 8-13. The resistance measured is related to sheet resistance ρ_s by

$$(8\text{-}23) \quad R = \frac{V}{I} = \rho_s \frac{L}{W}$$

The ratio L/W is the "number of squares" of the pattern. If contact resistance is a problem, two additional points on the body of the rectangle can be tapped to emulate the four-point probe technique. In this case, Eq. (8-23) is again used with L/W measured between the inner contacts.

8-3 ION IMPLANTATION

Ion implantation[10,11] is a method of introducing impurities in a controlled and precise manner into silicon wafers. As such, it serves the same function as a diffusion step. However, it has seen such widespread usage in MOS processes that it deserves discussion here in a separate section. Ion implantation was proposed relatively early in the history of IC technology. William Shockley had a basic patent issued in 1954, only seven years after the transistor was patented. Now it is used for such things as field V_T control, enhancement and depletion V_T adjust, CMOS tub or substrate doping, poly resistor doping, and source and drain doping. In fact it is safe to say that modern MOS circuits have close to 100 percent of their surface area covered with one implant or another. Even the backside of the wafer in some processes is implanted with argon for gettering.

The advantages of ion implantation over conventional thermal diffusion are

1. *Accurate Doping Control*   The amount of impurities introduced into the silicon wafer is controlled electronically and hence is very precise. It is easy to achieve better than 5 percent accuracy. Reproducibility is excellent, be it the resistance of implanted resistors or V_T of devices. Ion implantation is indispensable when the number of impurities to be introduced is small. For example, V_T adjust uses 10^{11} dopant atoms per square centimeter routinely, while chemical diffusion has a lower limit of 10^{13} cm^{-2}.

2. *Room Temperature Process*   Ion implantation is conducted at room temperature. The most significant consequence of this is that the photoresist pattern can directly mask the impurities. Photoresist can only withstand up to 150°C. Thus when diffusion is called for, the photoresist is used merely to etch an oxide layer, which in turn serves as the diffusion mask. The growth, etch, and subsequent stripping of the masking oxide layer can be eliminated with implantation.

3. *Implant through a Thin Layer*   Diffusion requires a surface free of oxide or nitride layers, while implants can penetrate such layers if they are thin enough. A good example of this is implantation through the gate oxide for threshold adjust. In fact it is a common practice to always implant through a thin oxide layer. This way the silicon surface stays protected from contaminants. Furthermore, breaking up the impinging ion beam improves uniformity.

4. *Profile Control*   Diffusion is constrained to have the peak concentration of impurities at the surface. Ion implantation offers the added degree of freedom of placing the concentration peak below the wafer surface. For example, certain short-channel devices have an implant below the

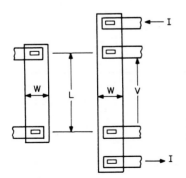

Figure 8-13 Sheet resistance measurement pattterns.

channel to raise the substrate doping locally to prevent punch-through between the source and drain.

Ion implantation is a process by which ions of a particular dopant are accelerated by an electric field and physically lodged into a silicon wafer by sheer momentum transfer. An ion implanter is illustrated in Fig. 8-14. The source of the ion can be gaseous or liquid (later vaporized). The source atoms are ionized in a plasma chamber and given an initial acceleration. A heavy magnet separates out the desired ion species by atomic weight (specifically, the charge-to-mass ratio). The ion beam is then focused and given a final acceleration. On the other end of the machine, wafers are placed to intercept the beam. Scanning of the beam across the wafer is accomplished by either electrostatically deflecting the beam in an *xy* raster scan or mechanically rotating a carousel or disk full of wafers in front of a steady beam. The number of ions impinging on the wafer is measured very precisely by means of a faraday cup. With the whole system under vacuum, the ions are positively charged and accelerated from -30 to -200 kV. For safety reasons, the wafer-handling end of the machine, which is the part that comes in contact with personnel, is kept at earth ground.

The design parameters of primary interest are the ion species, the voltage through which the ion is accelerated and the dose or number of ions impinging on the wafer surface per unit area. The choice of species is quite wide. It can consist of singly charged elemental ions, such as B^+, doubly charged ions such as B^{++}, or molecules such as BF_2. A singly charged ion accelerated through 100 kV would gain an energy of 100 keV. A doubly charged ion would have 200 keV. For the same energy, an ion with less mass would penetrate the silicon more deeply. The typical doses for implants used in MOS processing are given in Table 8-1.

The depth of penetration is quantified as the range R_p. Any single ion does not travel through the silicon in a straight line, but rather travels a

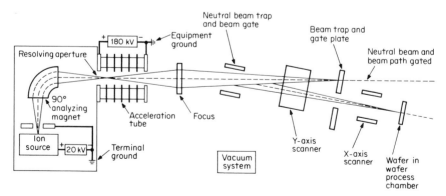

Figure 8-14   An ion implanter.

TABLE 8-1 Typical Dose for Implants Used in MOS Processing

Process step	Unique feature	Approx. range of dose, cm^{-2}
p-Tub	Dose control	10^{13}
Field implant	Dopant peak below surface	10^{13}
Threshold adjust	Control and low dose	10^{11}
Anti-punch-through	Profile control	10^{11}
Depletion implant	Control and low dose	10^{12}
Source and drain	Self-aligned	10^{15}
Getter	No phosphorus contamination	10^{15}

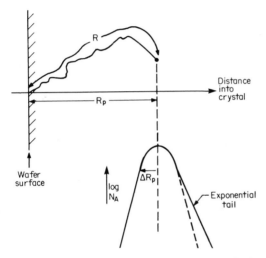

Figure 8-15 Definition of range and standard deviation for ion implantation.

random zigzag path from collisions with silicon nuclei before coming to rest. However, collectively, all the ions will come to rest with an average perpendicular distance from the surface of R_p as Fig. 8-15 shows. The distribution profile is, to a first approximation, gaussian with the standard deviation given by ΔR_p. A plot of R_p and ΔR_p as a function of energy is given in Fig. 8-16 for the three most common species, namely boron, phosphorus, and arsenic. Computer-generated tabulations of R_p and ΔR_p are also available[12] for a wide variety of ions and for background materials other than silicon. As the implant energy increases, both R_p and ΔR_p increase. This means

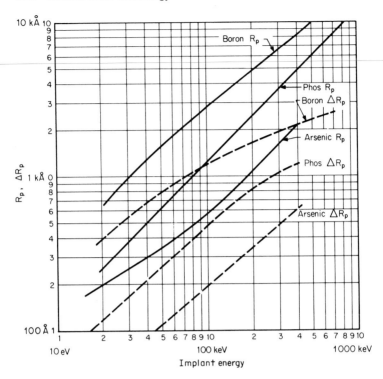

Figure 8-16 Range and standard deviation for ion implantation. R_P, solid; ΔR_P, dashed. *(After Refs. 10, 12).*

that for a given dose, as the implant gets deeper, it spreads out more and the peak concentration decreases.

The spatial distribution of implanted dopants can be modeled to first approximation by a gaussian distribution. However, boron has been found to have a significant "exponential tail" (Fig. 8-15) where a certain amount of dopant traveled farther than expected. This is attributed to the phenomenon of *channeling*. Silicon, being crystalline, has certain orientations where the atoms line up perfectly and the implanting ion sees only long empty channels. Furthermore, the repulsive forces of the silicon atoms guide the dopants to remain traveling down the channel. A small number of ions find their way into these channels and end up with longer R_p. Channeling is actually an undesirable effect because it is difficult to control. To prevent channeling, wafers are kept at 7° off normal with respect to the ion beam.

The gaussian distribution provides that the concentration of implanted ions be given by

$$(8\text{-}24) \quad N = N_{\text{peak}} \exp \left[-\frac{1}{2} \left(\frac{x - R_p}{\Delta R_p} \right)^2 \right]$$

The dose, equivalent to the area under the gaussian curve, is related to N_{peak} by

$$(8\text{-}25) \quad N_{peak} = \frac{D_s}{\Delta R_p \sqrt{2\pi}}$$

Example: If boron is implanted into 5×10^{15} cm^{-3} n-type silicon at 100 keV, and at a dose of 1×10^{13} cm^{-2}, what is the junction depth right after implant?

Solution: At 100 keV, from Table 8-2, one gets $R_p = 0.30$ μm, $\Delta R_p = 0.07$ μm.

$$(8\text{-}26) \quad N_{peak} = \frac{10^{13} \text{ cm}^{-2}}{(0.07 \times 10^{-4} \text{ cm})\sqrt{2\pi}}$$

$$= 5.7 \times 10^{17} \text{ cm}^{-3}$$

$$x_j = R_p \pm \sqrt{2}\,\Delta R_p \sqrt{\ln \frac{N_{peak}}{N_B}}$$

$$= 0.30 + \sqrt{2}(0.07) \sqrt{\ln \frac{5.7 \times 10^{17}}{5 \times 10^{15}}} = 0.52 \text{ } \mu\text{m}$$

It was pointed out earlier that one key advantage of ion implantation is the ability to introduce dopant through an intervening layer. The effect of such a layer is handled by converting its thickness to an equivalent silicon thickness as follows:

$$(8\text{-}27) \quad t_{eq} = t_{layer} \frac{R_p \text{ in silicon}}{R_p \text{ in layer}}$$

The range of the implant for materials other than silicon is given in Tables 8-2 to 8-4. This principle allows one to calculate the fraction of the implant dose that penetrates the layer and ends up in the silicon. That fraction f_{Si} is obtained by integrating the gaussian distribution from t_{eq} to infinity and dividing by the total dose. An error-function type of relationship results which can be plotted on a probability graph such as Fig. 8-17. The relationship is

$$(8\text{-}28) \quad f_{Si} = \frac{\displaystyle\int_{t_{eq}}^{\infty} N(x)\,dx}{D_s}$$

$$= \frac{\displaystyle\int_{t_{eq}}^{\infty} N_{peak} \exp\left[\frac{-1}{2}\left(\frac{x - R_p}{\Delta R_p}\right)^2\right]}{N_{peak}\,\Delta R_p \sqrt{2\pi}}$$

TABLE 8-2   Range and ΔR_p for Boron into Nonsilicon Substrate

Energy, keV	B in Si		B in SiO$_2$		B in Si$_3$N$_4$	
	Projected range, μm	Projected standard deviation, μm	Projected range, μm	Projected standard deviation, μm	Projected range, μm	Projected standard deviation, μm
10	0.0333	0.0171	0.0298	0.0143	0.0230	0.0111
20	0.0662	0.0283	0.0622	0.0252	0.0480	0.0196
30	0.0987	0.0371	0.0954	0.0342	0.0736	0.0267
40	0.1302	0.0443	0.1283	0.0418	0.0990	0.0326
50	0.1608	0.0504	0.1606	0.0483	0.1239	0.0377
60	0.1903	0.0556	0.1921	0.0540	0.1482	0.0422
70	0.2188	0.0601	0.2228	0.0590	0.1719	0.0461
80	0.2465	0.0641	0.2528	0.0634	0.1950	0.0496
90	0.2733	0.0677	0.2819	0.0674	0.2176	0.0527
100	0.2994	0.0710	0.3104	0.0710	0.2396	0.0555
110	0.3248	0.0739	0.3382	0.0743	0.2610	0.0581
120	0.3496	0.0766	0.3653	0.0774	0.2820	0.0605
130	0.3737	0.0790	0.3919	0.0801	0.3025	0.0627
140	0.3974	0.0813	0.4179	0.0827	0.3226	0.0647
150	0.4205	0.0834	0.4434	0.0851	0.3424	0.0666
160	0.4432	0.0854	0.4685	0.0874	0.3617	0.0684
170	0.4654	0.0872	0.4930	0.0895	0.3807	0.0700
180	0.4872	0.0890	0.5172	0.0914	0.3994	0.0716
190	0.5086	0.0906	0.5409	0.0933	0.4178	0.0731
200	0.5297	0.0921	0.5643	0.0951	0.4358	0.0744

Source: After Ref. 12.

$$= \tfrac{1}{2}\left[1 + \operatorname{erf}\left(\frac{R_p - t_{eq}}{\sqrt{2}\,\Delta R_p}\right)\right]$$

According to Fig. 8-17, if R_p is greater than t_{eq}, then more than 50 percent of the implant will reside in silicon. This is a quick check for reasonableness with regard to the sign for the argument of the error function.

Ion implantation can be masked by a film of photoresist or other materials if they are thick enough. Masking is considered effective if the $R_p + 3\Delta R_p$ point lies within the masking film. Figure 8-18 shows the film thickness required to mask the more common implants when performed at 100 keV. Data for several masking materials are included.

When an implanted ion comes to rest within the silicon, it leaves behind a track of atoms knocked from their lattice sites. In fact, under high-dose implants, these damage tracks merge to form an amorphous layer. It is therefore necessary to *anneal* the lattice disorder and radiation damage. When annealing is complete, the implanted ions will reside in substitutional sites and become electrically active. The silicon is restored to its high-quality preimplantation condition, manifesting high carrier bulk mobility, high lifetime, low sheet resistance, and low junction leakage. With low-dose implants, it is often sufficient to anneal at 600°C for half an hour in an inert ambient. But for high-dose implants, anneals at 900°C are required to achieve 100 percent activation. What is somewhat surprising is that implantations which result in a totally amorphous layer anneal completely at low temperatures of around 600°C, whereas boron, which creates less damage because of its low mass, requires temperatures as high as 900°C for proper anneal.

Conventional ion implant anneal is by thermal means, but increasing attention has been focused on using *laser annealing* instead.[13,14] Photon energy from the laser is absorbed by silicon that is very near the surface and converted to heat. The laser beam spot size is quite small, allowing the high temperature to be confined to a localized area. The wafer itself remains at ambient temperature, and there is negligible redistribution of dopants. Figure 8-19 shows a laser annealed doping profile to be essentially the same as

TABLE 8-3 Range and ΔR_p for Phosphorus into Nonsilicon Substrate

	P in Si		P in SiO$_2$		P in Si$_3$N$_4$	
Energy, keV	Projected range, μm	Projected standard deviation, μm	Projected range, μm	Projected standard deviation, μm	Projected range, μm	Projected standard deviation, μm
10	0.0139	0.0069	0.0108	0.0048	0.0084	0.0037
20	0.0253	0.0119	0.0199	0.0084	0.0154	0.0065
30	0.0368	0.0166	0.0292	0.0119	0.0226	0.0092
40	0.0486	0.0212	0.0388	0.0152	0.0300	0.0118
50	0.0607	0.0256	0.0486	0.0185	0.0376	0.0143
60	0.0730	0.0298	0.0586	0.0216	0.0453	0.0168
70	0.0855	0.0340	0.0688	0.0247	0.0532	0.0192
80	0.0981	0.0380	0.0792	0.0276	0.0612	0.0215
90	0.1109	0.0418	0.0896	0.0305	0.0693	0.0237
100	0.1238	0.0456	0.1002	0.0333	0.0774	0.0259
110	0.1367	0.0492	0.1108	0.0360	0.0856	0.0280
120	0.1497	0.0528	0.1215	0.0387	0.0939	0.0301
130	0.1627	0.0562	0.1322	0.0412	0.1022	0.0321
140	0.1757	0.0595	0.1429	0.0437	0.1105	0.0340
150	0.1888	0.0628	0.1537	0.0461	0.1188	0.0358
160	0.2019	0.0659	0.1644	0.0485	0.1271	0.0377
170	0.2149	0.0689	0.1752	0.0507	0.1354	0.0394
180	0.2279	0.0719	0.1859	0.0529	0.1437	0.0411
190	0.2409	0.0747	0.1966	0.0551	0.1520	0.0428
200	0.2539	0.0775	0.2073	0.0571	0.1602	0.0444

Source: After Ref. 12.

TABLE 8-4 Range and ΔR_p for Arsenic into Nonsilicon Substrate

Energy, keV	As in Si		As in SiO₂		As in Si₃N₄	
	Projected range, μm	Projected standard deviation, μm	Projected range, μm	Projected standard deviation, μm	Projected range, μm	Projected standard deviation, μm
10	0.0097	0.0036	0.0077	0.0026	0.0060	0.0020
20	0.0159	0.0059	0.0127	0.0043	0.0099	0.0033
30	0.0215	0.0080	0.0173	0.0057	0.0135	0.0045
40	0.0269	0.0099	0.0217	0.0072	0.0169	0.0056
50	0.0322	0.0118	0.0260	0.0085	0.0202	0.0066
60	0.0374	0.0136	0.0303	0.0099	0.0235	0.0077
70	0.0426	0.0154	0.0346	0.0112	0.0268	0.0087
80	0.0478	0.0172	0.0388	0.0125	0.0301	0.0097
90	0.0530	0.0189	0.0431	0.0138	0.0334	0.0108
100	0.0582	0.0207	0.0473	0.0151	0.0367	0.0118
110	0.0634	0.0224	0.0516	0.0164	0.0400	0.0127
120	0.0686	0.0241	0.0559	0.0176	0.0433	0.0137
130	0.0739	0.0258	0.0603	0.0189	0.0467	0.0147
140	0.0791	0.0275	0.0646	0.0201	0.0500	0.0157
150	0.0845	0.0292	0.0690	0.0214	0.0534	0.0167
160	0.0898	0.0308	0.0734	0.0226	0.0568	0.0176
170	0.0952	0.0325	0.0778	0.0239	0.0603	0.0186
180	0.1005	0.0341	0.0823	0.0251	0.0637	0.0195
190	0.1060	0.0358	0.0868	0.0263	0.0672	0.0205
200	0.1114	0.0374	0.0913	0.0275	0.0706	0.0214

Source: After Ref. 12.

the arsenic-implanted profile, whereas the thermally annealed case has significant redistribution.

The lasers used for annealing can be either continuous wave (CW) or pulsed.[15] One CW laser used successfully is an argon laser with wavelengths of 0.48 and 0.51 μm. The spot is on the order of 100 μm and is raster-scanned in a manner that gives a dwell time of 1 ms. Scan lines are adjusted to create 40 percent overlap. A pulsed laser, such as a Q-switched Nd:YAG (neodymium:yttrium aluminum garnet), produces a 50-ns pulse with a spot size of 5000 μm at a 1.06-μm wavelength. Although both types of lasers can be adjusted to have the same effective power density, as a matter of practice, the power density of pulsed lasers is high enough to cause localized melting of silicon. The dopant in that small volume of melt will redistribute uniformly. The silicon then recrystallizes extremely rapidly by means of liquid phase epitaxy, yielding a quality, defect-free crystal surface. A CW laser, meanwhile, is adjusted to recrystallize the wafer by means of solid-

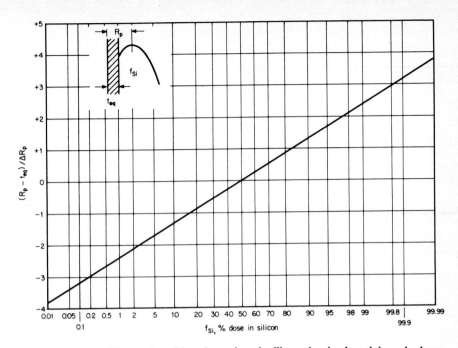

Figure 8-17 Fraction of dose that ends up in silicon when implanted through a layer.

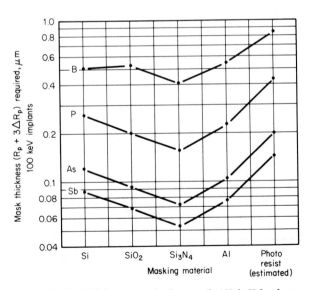

Figure 8-18 Thickness required to mask 100-keV implants. *(After Ref. 3.)*

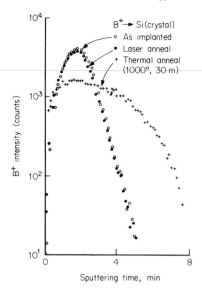

Figure 8-19 Doping profiles, As-implanted, laser-annealed, and thermally annealed. (*After Ref. 13.*)

phase epitaxy, also yielding a quality surface but without actually melting the silicon. The As-implanted profile is thus preserved.

Initial interest in the use of lasers in semiconductor processing came from the possibility of a direct replacement for thermal annealing. For example, in order to scale down junction depths, the NMOS process can switch the source and drain dopant from phosphorus to arsenic. But for the CMOS process, there is no such similar replacement for boron. Laser annealing opens up the means to activate fully such high-dose source and drain implants without disturbing the shallow profile. Lasers have since been found to be beneficial in other areas of MOS processing. Laser energy can recrystallize polysilicon films into larger grain sizes, resulting in a much lower sheet resistance. It also smoothes out the protuberances on such films, reducing poly to poly shorts when a second poly layer is deposited after first poly oxidation. It can replace thermal sources for sintering or alloying contacts to give lower contact resistances. Also, laser pulses, when applied to the back of a wafer, can create molten sinks for metallic impurities, resulting in an effective gettering procedure. Finally, lasers are being used to recrystallize polysilicon films into single crystals. The process, called *graphoepitaxy,* opens up the possibility of building different layers of circuitry on a wafer, i.e., integrated circuits in three dimensions.

8-4 DEPOSITION

Epitaxial growth of high-quality, high-resistivity films on a heavily doped substrate is finding increased use for very complex MOS ICs and for latch-up-free CMOS.[16] Epitaxial growth can be accomplished by the hydrogen reduction of silicon tetrachloride at very high temperature[17,18]:

$$
(8\text{-}29) \quad SiCl_4 + 2H_2 \text{ (gas)} \xrightarrow{1250°C} Si \text{ (solid)} + 4HCl \text{ (gas)}
$$

A second way is by pyrolysis (decomposition) of silane at somewhat lower temperature:

$$
(8\text{-}30) \quad SiH_4 \text{ (gas)} \xrightarrow{1000°C} Si \text{ (solid)} + 2H_2 \text{ (gas)}
$$

In both cases, the complete process starts with the use of low-defect substrates. For (111) material [but not for (100) material], the creation of nucleation sites on the wafer is enhanced by using a wafer that is 3 to 7° off a major axis to expose the edges of successive layers of crystal. Next is the introduction of dilute amounts of HCl in the hydrogen carrier gas to etch away the top 0.2 to 0.5 μm of silicon. This also acts to clean the wafer substrate. The wafers, which sit on silicon-carbide–coated graphite susceptors, are heated by radio frequency (RF) induction or high-intensity quartz light to start the chemical reactions. The equipment configurations (Fig. 8-20) may take the form of a rotating susceptor at the bottom of a bell jar, a long narrow boat in a diffusion tube, or a rotating barrel.

Epitaxial growth is actually a specific form of *chemical vapor deposition (CVD)* that requires the crystal structure of the substrate be extended into

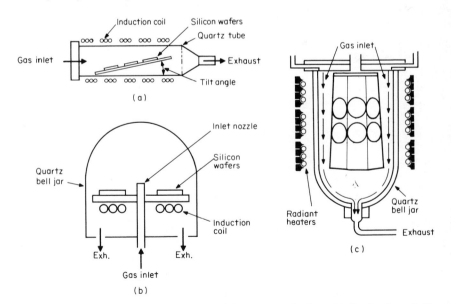

Figure 8-20 Different types of epitaxial reactors: (*a*) horizontal induction-heated, (*b*) vertical induction-heated, and (*c*) barrel radiantly heated. (*After Ref. 17.*)

the deposited layer. Regular CVD can be used to deposit polysilicon, silicon dioxide, or silicon nitride.[18,19] The deposition of polysilicon uses the same chemical reaction of Eq. (8-30), except that it takes place at 600 to 700°C. Polysilicon is usually deposited undoped and doped later in processing. Silicon dioxide is formed by

(8-31)
$$SiH_4 + 2O_2 \xrightarrow{450°C} SiO_2 + 2H_2O$$

The deposition takes place at 400 to 500°C and the film is sometimes referred to as LTO (low-temperature oxide). The oxide may be undoped or doped with arsenic, boron, or phosphorus by including arsine, diborane, or phosphene in the reaction. Deposited oxide is not as dense as thermal oxide, but heating it to 900°C or higher for 30 min "densifies" the film until the two are almost indistinguishable. Silicon nitride is formed by

(8-32)
$$3SiH_4 + 4NH_3 \xrightarrow{700°C} Si_3N_4 + 12H_2$$

Low-pressure CVD (LPCVD) is chemical vapor deposition performed at a pressure of 0.5 to 1 torr instead of the atmospheric value of 760 torr. At such low pressure, the diffusivity increases by three orders of magnitude.[20] This increase in mean free path of the gases allows an order of magnitude increase in the rate of transfer of reactant gases to, and by-products from, the substrate surface. A practical consequence is the ability to process wafers stacked on edges instead of lying flat as in conventional CVD. The resulting film exhibits better uniformity, better film integrity, and better step coverage and conformality. LPCVD can be used for deposition of polysilicon, silicon nitride, and silicon dioxide.

Aluminum is the most prevalent *metallization* material in MOS ICs. A film thickness of around 1 μm (10,000 Å) is usually required. The film is deposited by heating the aluminum source until it evaporates. The evaporant molecules strike the wafers and condense into a solid film. To improve uniformity, the wafers are made to travel on a planetary track with simultaneous rotation around several axes with respect to the source. To lower the latent heat of evaporation of aluminum to a workable level, the evaporation chamber must be in a vacuum. It is pumped down to 10^{-6} to 10^{-7} torr (often in stages, and using different types of pumps) before deposition can take place. Once evaporation starts, the pressure rises to 10^{-5} torr and is maintained at that value throughout the remainder of the cycle. Keeping the evaporation chamber in vacuum also eliminates contamination and increases the mean free path of the aluminum vapor.

The most common system of providing the evaporated aluminum is by focusing an electron beam (also called *E beam*) on a crucible of aluminum (Fig. 8-21a). Since only electrons come in contact with aluminum, the level of contamination can be very low. The E beam, which provides the direct power to vaporize the aluminum, is bent through 270° to reach the target.

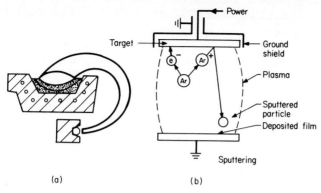

Figure 8-21 Two types of aluminum deposition systems: (*a*) E beam and (*b*) sputtering.

Because of the intensity of the electron beam used, there is radiation damage to the wafer which must be annealed out by subseqent process steps.

A common problem with aluminum film is its inability to conform to a step in the underlying oxide.[21] A hairline crack can develop right at the step. Heating the wafer to around 300°C during deposition often solves the problem. Another problem is *electromigration*.[22] In contact regions, silicon dissolves into aluminum until the 1 to 3% solid solubility is reached. Aluminum in turn replaces the vacated silicon, forming a metal spike that penetrates and shorts out the junction underneath. A second electromigration effect is the transport of aluminum by collision and momentum exchange with electrons in a current flow. Actual voids form on one end of a conduction stripe, while accumulation in the form of hillocks forms on the other end. These two effects are minimized by using aluminum with 1% silicon and lowering the current density.

A second metal deposition system is *sputtering* (Fig. 8-21*b*). An inert gas, usually argon, is brought to glow discharge in a vacuum by an electric field. The positive ions bombard the cathode, which is called the *target* and is coated with a solid layer of the metal to be deposited. By direct momentum transfer, the target material is dislodged and deposits on the wafers placed on the anode. The electric field can come from either dc or RF voltages (rectified by the plasma). Sputtering can be used to deposit practically any material by using the appropriate target, including nonconducting silicon dioxide, but the deposition rate is slow. To have a workable sputtering process, close attention must be paid to maximizing sputtering yield, defined as the number of dislodged atoms per bombarding ion.

8-5 PHOTOLITHOGRAPHY

Photolithography encompasses the complete process of transferring an image from a photographic mask to a resultant pattern on a wafer. It is a very labor intensive part of MOS IC fabrication because each wafer is processed individually and has to go through many substeps. At the heart of photolithography is photoresist, an organic polymer whose characteristics are drastically altered by exposure to light.

The most common type of photoresist (resist for short) is negative resist. Where it is exposed to strong ultraviolet (UV) light, the polymer cross-links, increases its molecular weight, and becomes impervious to solvents. The areas kept dark are dissolved away by developers. Negative resist is so called because the clear or opaque patterns produced on it are opposite the mask. Figure 8-22*a* illustrates this relationship. The other type of photoresist is positive resist. Whenever it is exposed to light, it is depolymerized and can be dissolved by developers. The unexposed areas remain impervious to solvents. Figure 8-22*b* shows that the clear or opaque patterns are identical to the mask, producing a positive image. For both types of resist, the solubility in the solvents changes abruptly within a narrow range of molecular weight, allowing excellent edge acuity. Positive resist can resolve finer lines and spaces, making it a favored choice for advanced high-density processes.

The photolithography process will be described below. First, for the resist to adhere properly, the wafer must be perfectly clean and moisture-free. A dehydration bake after cleaning is called for. The wafer is then placed on a chuck and held down by drawing a vacuum, and spun at several thousand revolutions per minute (r/min). At this point, adhesion promoters such as HMDS (hexamethyldisilazane) are sometimes applied. The resist is then dispensed, and excess resist will spin off. The final thickness is dependent on the spin speed and resist viscosity (in units of centipoise) and *not* on the amount of material dispensed. Figure 8-23 shows the film thickness as a function of spin speed for two representative photoresists. Note that positive photoresist tends to be thicker. The wafers are "soft-baked" (prebaked) at less than 100°C before exposure to drive off solvents in the photoresist. After exposure, the resist is developed. Wafers are then

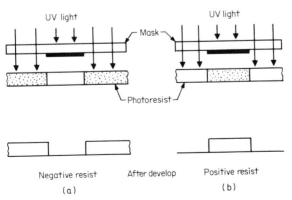

Figure 8-22 Negative and positive photoresist.

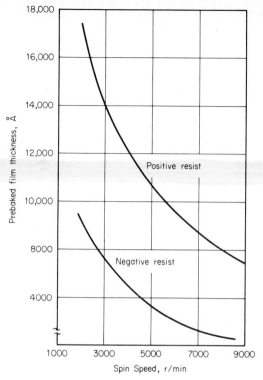

Figure 8-23 Photoresist thickness as a function of spin speed.

inspected individually under a microscope for overall quality and to measure critical dimensions (CD) to ensure that a certain line width or line spacing meets specifications. If all goes well, a "hard bake" (postbake) at slightly above 100°C is performed for maximum adhesion before subjecting the wafers to etchants. Because resists are organic materials, they can stand up to most chemicals used for etching polysilicon, oxide, or metal. Conversely, after etch, photoresist can be stripped with organic solvents such as acetone without harm to underlying layers.

There are three major types of exposure systems. The most basic system is *direct contact* printing. The aligner operator, using a microscope, aligns the mask with respect to patterns already on the wafer. The mask, emulsion side toward the wafer, is brought in direct contact with the wafer. Contact printing has the advantages of high resolution (down to 1-μm lines have been demonstrated), high throughput, and low cost. However, the direct contact between mask and wafer causes the mask to deteriorate with use. A photographic emulsion type of mask is limited to 10 to 15 exposures. Replacing the emulsion with chrome extends the useful life to 150 exposures, but the defects it creates on the photoresist pattern makes it impractical for large-scale integrated circuits.

The *projection printing* system projects the mask's pattern through reflective lens elements (instead of refracting lenses) to the wafer without the two coming into contact.[27] To improve resolution, the image field size is reduced by focusing the light through an arc-shaped slit 1 mm wide. The mask moves past the slit with the wafer moving in unison, so the whole mask is exposed in a single scan (see Fig. 8-24a). The mask can easily last up to 150,000 exposures; thus it is economically feasible to use a directly stepped master plate as the mask. But savings gained in mask cost by using projection printing are overshadowed by the improvement in yield due to reduced defects.

As the resolution and alignment accuracy of exposure systems improve, certain limitations become more evident. One is the warpage of the wafer due to heat cycles in processing; another is the mismatch in the coefficient of expansion between the mask and the wafer. One solution to both problems is the *direct step on wafer* (DSW) aligner. A 10X reticle (pattern for

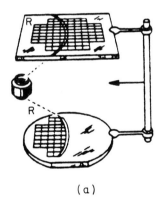

(a)

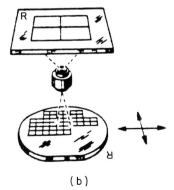

(b)

Figure 8-24 Two types of aligner and exposure systems: (a) projection and (b) direct step. (*After Ref. 27. Reprinted with permission of Technical Publishing, a Company of Dunn & Bradstreet.*)

one die) is focused down on an individual die location as shown in Fig. 8-24*b*, and the exposure stepped and repeated across the wafer. Prior to each exposure, minor adjustments in focus and alignment can be carried out electronically. The use of a 10X reticle allows minor defects and dust particles to be reduced in size by a factor of 10, beyond the resolution limit of the photoresist. On the other hand, any major defect on the reticle will be replicated on every die of the wafer, producing zero yield. For this reason, reticles must be inspected meticulously before being qualified for use. DSWs have high capitalization costs and slow throughput, but their higher resolution and greater alignment accuracy (overlay) make them indispensable for large-scale integrated circuits.

Instead of exposing photoresist through a mask, an electron beam can be used to "write" directly on a wafer. Such E-beam pattern generators evolved from scanning electron microscopes. The exposure is accomplished either by a raster scan, where the beam is fixed in position and the wafer moves in an *xy* raster scanning motion, or by a vector scan, where the wafer is fixed and the beam moves to where exposure is needed. For masks whose pattern density is low, vector scan increases throughput significantly. Further increase in throughput is possible if the beam size *and* shape are changed as needed during exposure. Because of the long exposure time, E-beam machines find their initial applications in producing integrated circuit masks. They also find applications in research and development where the attraction lies in the fact that changes in the circuit are accomplished by simply editing the software. Disadvantages of E-beam machines are their high cost and slow throughput. Exposure times vary from 15 min to 1 h per layer. More sensitive and faster resists are under development.

Since resolution is improved by using shorter wavelengths, orders of magnitude improvement is possible by going from an optical to an x-ray source. Some of the materials used as sources are Pd (4.4 Å), Mo (5.4 Å), W (7.4 Å), Al (8.3 Å), and Cu (13.1 Å). The mask consists of a thin Si membrane a few micrometers thick with gold overlay, which is opaque to x-ray, as the pattern. Research is continuing in improving the sensitivity of the resist to reduce exposure time.

The availability of successive generations of exposure and etching (next section) equipment allows the relentless shrinking of feature size on an integrated circuit. Figure 8-25 illustrates the point, and shows that in the first half of the 1980s, the bulk of routine production will be in the 3- to 4-μm range.[23]

8-6 ETCH

Photolithography and etch are two closely coupled processes. Genuine progress is made not only if each is individually workable, but also if the two are compatible. The ultimate criterion is that the photoresist be able to

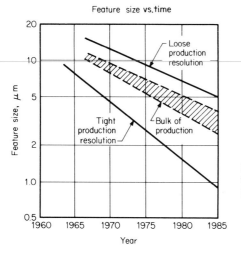

Figure 8-25 Feature size vs. time as achievable in production. (*After Ref. 23.*)

retain its pattern integrity while being resistant to the etchant used. The etching process itself will be examined in this section.

The most common etching process is the wet (chemical) etch where a batch of wafers with patterned resist is immersed in a temperature-controlled etchant for a fixed period of time. The amount of exposed material removed is a linear function of time, with the etch rate being determined by the strength of the etchant, the temperature, and the material being etched. Table 8-5 shows the etch rates for several etchants commonly used. Oxides are etched by hydrofluoric acids. HF acid is often diluted with ammonium fluoride (NH_4F) to slow down and stabilize the etch rate to a manageable level. The mixture is also referred to as buffered HF. The buffer agent regulates the chemical activity of the acid by keeping the number of reactive ions constant even as the solution is used.

The etch rate of the underlying layer should be much lower than the material being etched. This provides a natural etch stop in case the etching continues longer than needed. A selectivity of 10:1 or higher is the norm.

Silicon nitride is etched by hot phosphoric acid, H_3PO_4, polysilicon is etched by a solution of nitric acid, acetic acid, and HF, while aluminum is etched by a phosphoric acid mixture.

Chemical etch is isotropic, that is, as the chemical etches down through the material, it also etches laterally by approximately the same amount. A thick layer then produces a large "undercut." Overetching tends to exacerbate the situation since the lateral etching continues unabated even though there is a vertical etch stop. Furthermore, the photoresist tends to lift at the etch opening. All of the above cause the etched opening to be larger than that patterned on the photoresist. For high-density, very large scale integrated circuits, an anisotropic etch is necessary to provide vertical etching with nearly zero lateral undercutting from the photoresist opening. This capability can be provided by plasma etching.

Plasma etching is an alternative method of etching that involves gases ionized and rendered chemically active by an RF generated plasma. The reaction by-products are volatile and are carried away by the gas flow. Since the process does not involve wet chemicals, it is sometimes referred to as *dry etching*. Several gases containing chlorine or fluorine, notably carbon tetrachloride, CCL_4, and carbon tetrafluoride, CF_4, can be tailored for etching either polysilicon, silicon nitride, silicon dioxide, or aluminum (chlorine only). Adding 5 to 10% O_2 increases the etch rate significantly for some of the materials. On the other hand, pure oxygen is used for plasma stripping (removal) of photoresist.

Early plasma reactors were barrel type where a perforated metal cylinder confines the plasma to its exterior. The batch of vertically stacked wafers is placed inside the cylinder, shielded from UV radiation emitted by the plasma. The active species are electrically neutral, so the reaction is entirely chemical. This type of plasma etching is still isotropic, similar to wet etching.

The parallel plate type of plasma reactor places the wafer flat on a platen that serves as one of the RF electrodes. A net dc voltage develops that drives the active species toward the wafers. This mixture of physical and chemical etching is referred to as *reactive ion etching* (RIE)[24] and results in anisotropic etching, i.e., the vertical etch rate is substantially higher than the lateral etch rate.

Plasma etching requires significant process characterization. Variables such as pressure, gas flow rate, gas mixture, and RF power must be optimized to produce the desired etch rate, selectivity, amount of anisotropy

TABLE 8-5 **Etch Rate for Various Etchants**

	Conc. HF	H_2O:HF 10:1	NH_4F:HF 10:1
OXIDE ETCH			
Thermal SiO_2	300	5.3	14
Deposited SiO_2	950	35	30

Å/s

AL ETCH
16:2:1:1 H_3PO_4:H_2O:HAc:HNO_3
$\qquad$ Etch rate = 2000 Å/min

NITRIDE ETCH
180°C H_3PO_4
$\qquad$ Etch rate = 65 Å/min.

(perfectly vertical walls are not always desirable), and etching uniformity. In addition, not all photoresists can withstand plasma processing.

The third type of plasma-assisted etching, where the etching takes place entirely by physical momentum transfer, is exemplified by ion milling.[25] Argon is ionized in a plasma, and positive ions are extracted and accelerated up to 1000 keV in a collimated beam. A negatively charged grid neutralizes the ions before reaching the target. Ion milling can be used to etch practically any material and thus finds such applications as in magnetic bubble memory fabrication. It is used in etching fine permalloy patterns for which no chemical etch has been found suitable. Ion milling also finds applications in metal etching where photoresist defects such as poor adhesion or slight bridging of adjacent photoresist lines are overcome with the direct milling action.

8-7 CLEAN

Both the wafer and its environment must be absolutely clean during wafer fabrication. Contamination control must be considered an integral part of wafer processing. There are three major areas for eradicating contamination: chemical cleaning of the wafer, using ultrapure water, and lowering the particle count in the air.

Wafer Cleaning

Cleaning of the wafer is done prior to every processing step and constitutes up to half the overall wafer processing activity. Cleaning is particularly important prior to high-temperature diffusion, oxidation, and deposition steps. Contamination by mobile alkali ions such as sodium and heavy metals such as gold and copper is particularly harmful. There are many cleaning procedures and most are company proprietary, but they tend to follow a common series of steps.

1. ***Degreasing to Remove Wax and Oils*** Chemicals such as 1,1,1-trichloroethane can be used, followed by rinses in acetone and alcohol, which are both completely volatile. This step is usually only needed on a raw wafer.

2. ***Chemically Oxidizing Away Organic Materials Such As Leftover Photoresist*** Chemicals such as hot sulfuric acid (H_2SO_4) are effective. Another effective solution is part of the "RCA clean."[26] It uses a high pH hydrogen peroxide solution mixed from 5:1:1 parts by volume of $H_2O:H_2O_2:NH_4OH$.

3. ***Removal of Heavy Metals*** The second RCA clean solution is a low-pH hydrogen peroxide solution containing 6:1:1 $H_2O:H_2O_2:HCl$ by

volume. This solution dissolves metals from the silicon wafer surface and forms soluble complexes to prevent replating from the solution.

4. ***Removal of the Top Layer of Oxide*** A brief dip in dilute HF removes the top layer of SiO_2 and any ionic contaminants that may be chemically adsorbed.

5. ***Rinse and Dry*** Rinsing with ultrapure water removes any residual acid. The wafers are then dried by blowing with filtered nitrogen or filtered air.

Water

Since water is usually the final rinsing solution, it must be as pure as possible. Only deionized water (referred to as DI water) will satisfy the requirements. It has a resistivity of 15 to 18 $M\Omega \cdot cm$ and contains fewer than 10 to 100 particulates or living organisms per cubic centimeter. Purification begins by chlorinating the incoming feed water. Then the water passes through a series of filters, namely, sand, activated charcoal, and diatomaceous earth, each one removing the next smaller particles. The actual deionization begins when water is passed through activated resins, and ion exchange takes place to remove anions and cations from strong acids and bases. Ultraviolet light is then used to control bacteria growth.

The water can be distributed throughout the facility by using inert plastic pipes. A double piping method keeps the water circulating even when not in use. A small final filter is placed near the point of use. Water treatment systems range in size from a few cubic feet to entire buildings.

Since the late 1970s, deionization systems have gradually been enhanced by addition of reverse osmosis systems. The various mechanical filters are replaced by a special membrane which allows water to flow through but not the dissolved or suspended particles. This filtration method is more effective and reduces the frequency with which the ion exchangers must be recharged.

Air

Airborne particles must be kept under constant control. The first line of defense is to keep wafers in covered containers, and whenever out of the box, they must be kept under laminar flow hoods. These are work stations that take the room air, pass it through a *h*igh-*e*fficiency *p*articulate *a*ir, or HEPA, filter, and blow it in a laminar flow pattern down on the work surface. The laminar flow provides a sweeping action across the work surface and prevents turbulent regions that may trap particles.

The laminar flow concept can be expanded to turn the whole work facility into a clean room. Air is blown through HEPA filters overhead, flows downward, and is exhausted near the floor. Modern MOS processing requires a Class 100 rating, meaning an environment with a maximum of 100 particles with a size larger than 0.5 μm in 1 ft^3. Other particle sizes scale in inverse proportion, e.g., no more than 10 particles per cubic foot larger than 5.0 μm. A Class 1000 room allows an order of magnitude greater particle count. In a clean environment, personnel become a major source of particles. There is thus a need for dust-confining jumpsuits, air showers at room entrances, hoods, lint-free paper, and other clean room practices.

REFERENCES

1. D. J. Hamilton and W. G. Howard, *Basic Integrated Circuit Engineering*, Chap. 2, McGraw-Hill, New York, 1975.

2. B. Deal and A. Grove, "General relationship for the thermal oxidation of silicon," *J. Appl. Phys.*, vol. 36, no. 12, pp. 3770–3778, December 1965.

3. O. D. Trapp, R. A. Blanchard, L. J. Lopp, and T. I. Kamins, *Semiconductor Technology Handbook*, Chaps. 4, 7, and 9, Technology Associates, Portola Valley, Calif., 1982. (*Note:* This handbook contains many charts and graphs that are useful in semiconductor processing and process control.)

4. R. J. Kriegler, "The role of HCl in the passivation of MOS structures," *Thin Solid Films*, vol. 13, pp. 11–14, 1972.

5. E. J. Janssens and G. J. Declerck, "The use of 1.1.1.-trichloroethane as an optimized additive to improve the silicon thermal oxidation technology," *J. Electrochem. Soc.*, vol. 125, no. 10, pp. 1696–1703, October 1978.

6. A. S. Grove, *Physics and Technology of Semiconductor Devices*, Chap. 3, Wiley, New York, 1967.

7. J. R. Ligenza, "Oxidation of silicon in high pressure steam," *J. Electrochem. Soc.*, vol. 109, pp. 73–76, February 1962.

8. D. R. Craven and J. B. Stimmell, "The silicon oxidation process—including high pressure oxidation," *Semiconductor Int.*, pp. 59–74, June 1981.

9. N. Tsubouchi et al., "The applications of the high-pressure oxidation process to the fabrication of MOS LSI," *IEEE Trans. Electr. Dev.*, vol. ED-26, no. 4, pp. 618–622, April 1979.

10. D. H. Lee and J. W. Mayer, "Ion implanted semiconductor devices," *Proc. IEEE*, pp. 1241–1255, September 1974.

11. J. F. Gibbons, "Ion implantation in semiconductors," *Proc. IEEE*, Part I, vol. 56, pp. 295–319, March 1968; Part II, vol. 60, pp. 1062–1096, September 1972.

12. J. Gibbons et al., *Projected Range Statistics*, 2d ed., Dowden, Hutchinson & Ross, Stroudsburg, Pa., 1975.

13. C. W. White et al., "Laser annealing of ion implanted semiconductors," *Science,* vol. 204, no. 4392, pp. 461–468, May 4, 1979.

14. M. Koyanagi et al., "Short channel MOSFETs fabricated by self-aligned ion implantation and laser annealing," *Appl. Phys. Lett.,* vol. 35, no. 8, pp. 621–623, Oct. 15, 1979.

15. J. F. Ready and B. T. McClure, "Laser annealing," *Semiconductor Int.,* pp. 93–112, November 1981.

16. G. R. Srinivasan, "Silicon epitaxy for high performance integrated circuits," *Solid-State Tech.,* pp. 101–110, November 1981.

17. R. W. Atherton, "Fundamentals of silicon epitaxy," *Semiconductor Int.,* pp. 117–130, November 1981.

18. J. L. Vossen and W. Kern (eds.) *Thin Film Processes,* Part III-2, Academic, New York, 1978.

19. P. E. Gise and R. Blanchard, *Semiconductor and Integrated Circuit Fabrication Technique,* Chaps. 5, 6, 10, and 12, Reston, Reston, Va., 1979.

20. W. A. Brown and T. I. Kamins, "An analysis of LPCVD system parameters for polysilicon, silicon nitride, and silicon dioxide deposition," *Solid-State Tech.,* pp. 51–84, July 1979.

21. A. J. Learn, "Evolution and current status of aluminum metallization," *J. Electrochem. Soc.,* vol. 123, no. 6, pp. 894–906, June 1976.

22. J. R. Black, "Electromigration—A brief survey and some recent results," *IEEE Trans. Electr. Dev.,* pp. 338–347, April 1969.

23. M. H. Eklund and G. Landrum, "1982 forecast on processing," *Semiconductor Int.,* pp. 43–55, January 1982.

24. L. M. Ephrath, "Reactive ion etching for VLSI," *IEEE Trans. Electr. Dev.,* vol. ED-28, no. 11, pp. 1315–1319, November 1981.

25. D. D. Robertson, "Advances in ion beam milling," *Solid-State Tech.,* pp. 57–60, December 1978.

26. W. Kern and D. A. Puotinen, "Cleaning solutions based on hydrogen peroxide for use in silicon semiconductor technology," *RCA Review,* pp. 187–206, June 1970.

27. R. K. Watts and J. H. Bruning, "A review of fine-line lithographic techniques: Present and future," *Solid-State Tech.,* pp. 99–105. May 1981.

PROBLEMS

1. A wafer is subjected to dry O_2 oxidation at 1000°C for 30 min, then O_2 + 5% HCl oxidation at 1000°C for 45 min. What is the final thickness? [Assume (100) material.]

2. The *p*-tub diffusion areas in CMOS are etched through 8000 Å of oxide. The wafer is then submitted to a steam oxidation cycle for 60 min at 1000°C. All oxide is stripped, and a step in the silicon can be seen that is used for subsequent alignment. How thick is that silicon step? (See Fig. 8-26.)

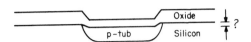

Figure 8-26 **Prob. 2.**

3. (a) A wafer sees a boron predeposition cycle at 950°C for 180 min. With a background concentration of 5×10^{15} cm^{-3}, find junction depth x_j.
   (b) The same wafer is next subjected to a drive-in cycle for 4 h at 1100°C. Find the new x_j.

4. Graph the dopant distributions for three phosphorus implants with a dose of 1×10^{15} cm^{-2} at (a) 30 keV, (b) 60 keV, and (c) 300 keV.

5. There are ion implanters specifically designed to handle high-dose implants. They have provisions for good heat sinking to prevent wafer temperature rise. However, they usually have only a limited voltage capability. If one such predeposition machine has a maximum of 80 keV and a boron source and drain implant has to pass through 900 Å of oxide, what fraction of the implant dose ends up in the silicon?

6. List the advantages and the disadvantages of contact printers, projection printers, direct step on wafers, E beam, and x-ray photolithography equipment.

7. A 1000-Å layer of SiO$_2$ is deposited over a 500-Å layer of thermal SiO$_2$. If the structure is etched by a 10:1 H$_2$O:HF solution, plot the total oxide thickness as a function of time throughout the complete etching step.

9

MOS Digital IC Design

9-1   BUILDING BLOCKS FOR MOS DIGITAL IC

A digital integrated circuit contains logic gates as its building blocks. In the case of MOS digital ICs, these logic gates are formed exclusively with MOS transistors, i.e., there are no resistor loads. This is one key reason why MOS circuits are denser than bipolars, since resistors occupy more area than equivalent transistors.

Three basic logic gates are shown in Fig. 9-1, namely, an *inverter,* a *NAND* gate, and a *NOR* gate. The associated logic symbol and truth table for each are also shown in the figure. A truth table lists the output states for all the possible combinations of input states.

An inverter functions by having the output always at an opposite logic level to the input. When the input is 0 (low voltage, or near ground) then the driver or pulldown device, to which the input is applied, turns off. The output node is charged to a 1 (high voltage, or near V_{DD}) through the load or pullup device. On the other hand, when the input is a 1, the driver device turns on, and the relative gain of the two transistors is designed such that the voltage dividing action of the two on-transistors will result in a 0 at the output.

By going through the truth table in conjunction with the circuit implementation, one can verify that the NAND gate's output will be a 0 *only* if *both* inputs are 1. Similarly, the NOR gate's output will be a 0 if *either* one or *both* inputs are 1. It is interesting to note that if one of the NAND gate's inputs is a 0, then changing the other input has no effect. That one input is thus acting as an enabling input, allowing an output transition only if it is a 1. When enabled, the logic gate is functioning as an inverter. A similar observation applies for a NOR gate when it is enabled by a 0.

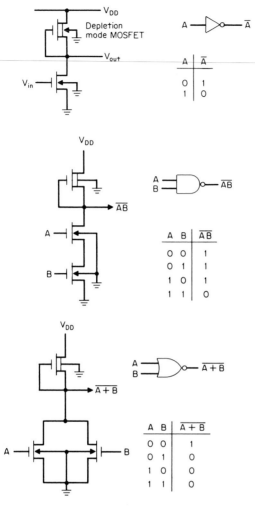

Figure 9-1 Three basic logic gates—inverter, NAND, and NOR—their logic symbols, and truth tables.

Note in Fig. 9-1 that the substrates of all devices are common and are connected to ground. This imposes a source-to-substrate bias on the load device when the output mode rises from ground.

The analysis of the above three logic building blocks reduces to that of a single inverter. If more than one driver device is on, they can be treated as one equivalent device in question. All others are off and out of the picture. Therefore an inverter deserves a closer analysis. It can conveniently be divided into dc and transient analysis.

9-2 INVERTER DC ANALYSIS

An inverter consists of a driver device in series with a load device. Four ways to implement an inverter are demonstrated in Fig. 9-2, depending on what kind of load device is used. A fifth kind is the CMOS inverter which is discussed in Chap. 11. The first method uses a resistor as the load; the second, an enhancement-mode transistor with its gate tied to the drain (saturated mode); the third, an enhancement-mode device whose gate is tied to a higher potential than its drain, thus keeping it in the triode region; and the fourth, a depletion-mode device whose gate is tied to its source.

Resistor Load[1]

The basic inverter with a passive linear resistor as load device is shown in Fig. 9-3. If one superimposes the load lines of various load resistors on the drain characteristics of the driver device, one obtains Fig. 9-4. As the input gate voltage changes from a low to maximum, the output V_{DS} is constrained to swing from maximum (in this case $V_{DD} = 12$ V) to minimum, along a single load line. If the output voltage is plotted as a function of input voltage, the transfer characteristics of Fig. 9-5 are obtained. When the input is below threshold (in this case, $V_T = 3.5$ V), the driver device is off and the output is at full supply voltage. When the input voltage increases to a 1 level, the output decreases to a 0. The higher the load resistance, the lower the output

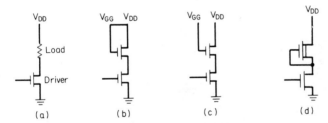

Figure 9-2 MOS inverter using four basic types of loads. (*a*) Linear resistance; (*b*) saturated MOS, $V_{GG} = V_{DD}$; (*c*) nonsaturated MOS, $V_{GG} - V_T \geq V_{DD}$; (*d*) depletion MOS.

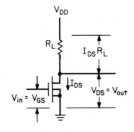

Figure 9-3 Inverter with passive resistor as load.

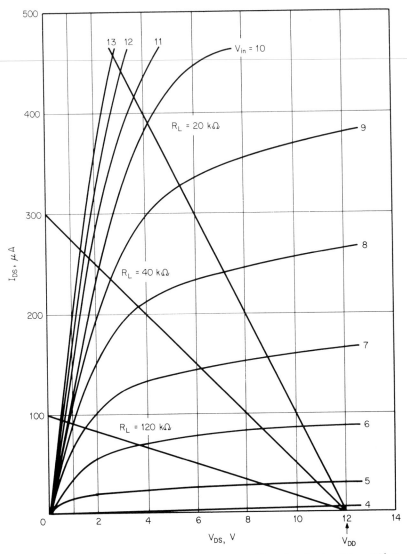

Figure 9-4 Drain characteristics of MOS inverter with various load resistors. (*After Ref. 1.*)

voltage for a 1 input since the output voltage is the result of voltage division between the load resistance and the on-resistance of the driver device.

Saturation Enhancement Transistor Load

When a saturation enhancement transistor is used as the load, the gain (W/L ratio) of the driver device is often made 10 to 20 times larger than the gain of the load. This results in a characteristic that looks like Fig. 9-6 when

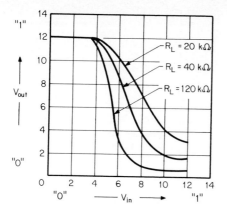

Figure 9-5 Transfer characteristics of MOS inverter with various load resistors. *(After Ref. 1.)*

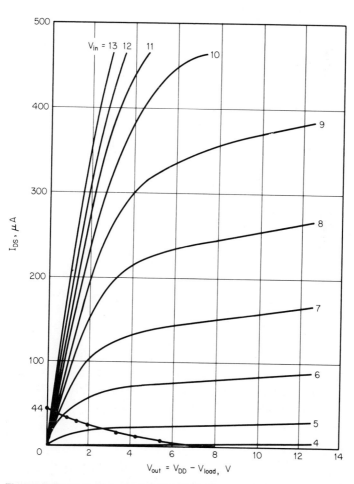

Figure 9-6 Drain characteristics of MOS driver device with load line of saturated load device. *(After Ref. 1.)*

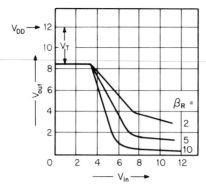

Figure 9-7 Transfer characteristics of inverters with saturated load device. *(After Ref. 1.)*

superimposed with its load line. The load line for the saturated transistor is nonlinear, but this is of no consequence in a digital circuit. The load line is in fact a square law type, corresponding to the two terminal characteristics of the saturated device.

The transfer characteristics are shown in Fig. 9-7. The single distinguishing feature is that the output no longer goes to the full V_{DD} value but is one threshold drop below. In fact, the output will be lower by more than one threshold drop because the load device has its source reverse-biased with respect to the substrate, increasing the threshold voltage by an amount ΔV_T because of the back-gate bias effect discussed earlier. Shown also is the fact that when the gain ratio (beta ratio) is made larger, the output voltage becomes lower, where

$$(9\text{-}1) \quad \beta_R = \frac{W/L \text{ of driver}}{W/L \text{ of load}}$$

Triode Enhancement Transistor Load

If the gate voltage of the load device is tied to a more positive voltage than V_{DD}, then the load device (Fig. 9-2c) is always constrained to be in the triode region. This is illustrated in the *VI* curve of the load device in Fig. 9-8. The case for saturated load device is also displayed for comparison.

The transfer characteristics are shown in Fig. 9-9. When compared to the saturated load case, one sees that the output can once again reach V_{DD}. Since the load device has a higher on-resistance, the output voltage for a given 1 input is higher than the saturated load case. A higher beta ratio is needed to achieve the same low output.

Depletion-Mode Load

With the advent of ion implantation, depletion-mode devices can be fabricated on the same wafer as enhancement-mode devices. The transfer characteristics of such inverters are shown in Fig. 9-10. Again, note that the output voltage can go all the way to V_{DD} since the load device is always on, even with its gate tied to the source. Another advantage of this form of inverter lies in its improvement in switching speed and will be evident in the next section on transient analysis.

The procedure for deriving the transfer characteristics of an inverter will now be outlined. Details of the derivation are available in the literature.[2,3,4] As the input increases from 0 V to V_T, the driver device will be off. The output will stay unchanged at a maximum value. Once the input exceeds V_T, then the inverter enters a linear, high-gain region where the driver device is in saturation. Finally, the gain drops off rapidly as the input is further increased and the driver device enters the triode region. At this point, the driver device is turned on "hard" and the full power supply voltage is effectively dropped across the load device. The current drawn by the inverter is maximum under this condition and is determined solely by the size of the load device.

Under dc conditions, at any given output voltage, the current flowing through the load device has to flow through the driver device too. By selecting the appropriate expression for the driver device current (triode or sat-

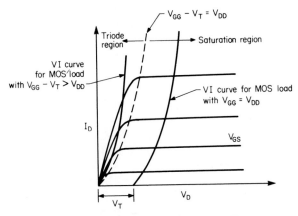

Figure 9-8 Drain characteristics of MOS load device with superimposed two-terminal *VI* curve.

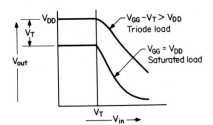

Figure 9-9 Transfer characteristics of inverters with triode and saturated loads.

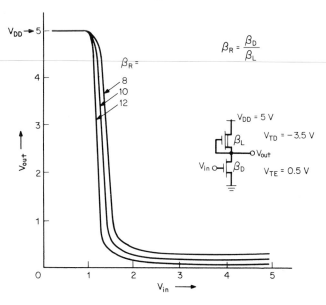

Figure 9-10 Transfer characteristics of inverters with depletion-load devices.

uration) and equating it to a similar expression for the load current, a relationship between V_{in} and V_{out} can be derived. This is the transfer curve. A family of curves will result for different values of β_R. For more accurate results, a new value of the load device threshold voltage must be calculated for every new V_{out} value obtained to take into account back-gate bias.

An inverter must be designed to have enough *noise margin* to reject noise spikes that couple to input nodes from clock or other signal lines by means of parasitic capacitances. Noise margin can be graphically determined from the dc transfer characteristics such as shown in Fig. 9-11. It is defined as the difference in input voltage between the operating point and the unity gain point on the curve. Separate noise margins are defined for 0 and 1 inputs. With this definition, noise spikes smaller than the noise margin will be attenuated in passing through the inverter. A less conservative approach would define noise margin as the voltage difference between the operating point and the intersection with the $V_{in} = V_{out}$ line. The argument for it is that the output does not change state unless noise spikes cause the input to actually cross the $V_{in} = V_{out}$ point. An inverter must be designed with large enough β_R that a 0 output will have low enough voltage to give good noise margin for the next inverter it drives (see Prob. 1).

For a NAND gate to provide proper 0 output level when both its inputs are on, the two driver devices in series must have the same on-resistance as a single device by itself. This is accomplished by doubling their W/L ratio.

9-3 INVERTER TRANSIENT ANALYSIS

An inverter must be designed from a dc point of view to provide the proper output levels to switch the next logic gate. In addition, it must change state within a specified period of time. This is the subject of transient analysis.

An equivalent circuit for transient analysis is given in Fig. 9-12. The output load is represented solely by a single capacitor. The input node responds instantaneously to a square wave excitation, but the output response is limited by the inverter's ability to charge or discharge the output capacitance.

In the first transition, when the input changes from a 1 to a 0, the driver device immediately shuts off. The output node then rises toward V_{DD} with a certain *risetime*. The risetime is determined solely by the load device and the output capacitance. With a saturated load ($V_{GG} = V_{DD}$) the load device shuts off before the output charges up fully, leaving it at one V_T drop below V_{DD}. Figure 9-13 illustrates the charge-up characteristics. With triode loads ($V_{GG} - V_T > V_{DD}$) or depletion loads, the load device stays on and the output charges fully to the V_{DD} level. In all cases, however, the current flow available through the load device decreases as the output voltage rises, slow-

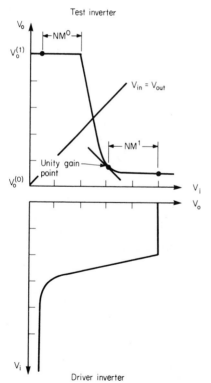

Figure 9-11 Noise margin measured from transfer characteristics.

ing down the charge-up process. The result resembles an RC exponential charge-up. Back-gate bias effect slows down the response from the ideal case even further.

In the second transition of Fig. 9-12, when the input charges from a 0 to a 1, the driver device turns on to discharge the capacitor to ground. To first approximation, the load device is ignored and the capacitor discharges through the driver device. The approximation is valid since the load device current is a small fraction of the total current the driver can supply for discharging the capacitor. The driver device passes from saturation region to triode region as the output voltage (its drain voltage) falls from a 1 to a 0 level. A typical response is given by Fig. 9-14. In comparing with Fig. 9-13, one should note that the *falltime* is normally several times faster than the risetime due to the fact that the driver device gain β_D is typically 10 times greater than the load device gain β_L so its corresponding time constant is much shorter.

It is very common to use computer simulation (see Chap. 12) to generate the transfer characteristics and the rise- and falltime transient response. Computer simulation allows the use of a more precise device model than has been presented thus far. Circuit simulation is at times a requirement, such as in transient analysis where the input waveform is allowed to have a finite

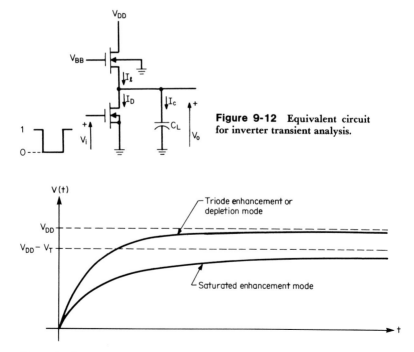

Figure 9-12 Equivalent circuit for inverter transient analysis.

Figure 9-13 Output risetime characteristics.

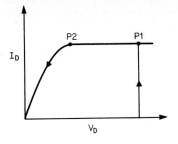

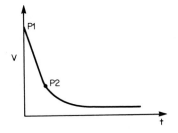

Figure 9-14 Output falltime characteristics.

rise- and falltime. In such a case, the propagation delay is also of interest. Propagation delay is the delay time between input and output waveforms and is measured between the 50 percent points. Rather than making a computer run for each situation, it is good engineering practice to perform a series of computer simulations and use the results to construct design curves such as shown in Fig. 9-15. Subsequently, for conditions that fall within the range of the graph, the answer can be obtained graphically. Figure 9-15 shows that the propagation delay when output is falling (t_{pd}^-) is determined by the driver device, and propagation delay when output is rising (t_{pd}^+) is determined by load device.

9-4 MOS LOGIC CIRCUITS

Combined AND-OR Functions

Aside from straightforward implementation of logic circuits with inverters, NAND and NOR gates, MOS circuits offer the flexibility of combining more than one level of logic. Figure 9-16 shows two examples of implementation of two levels of logic with only four devices: one load device and one driver device for each of the inputs. Such technique is efficient from the point of view of layout area, power dissipation, and propagation delay.

RS Flip Flops

A very useful and basic circuit configuration is a flip flop, which is used to store the status of a logic variable having one of two possible logic states. A

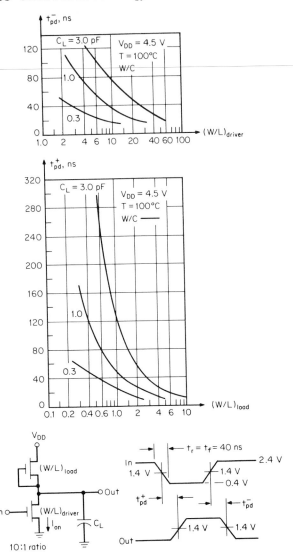

Figure 9-15 Propagation delays of inverters.

specific type of flip flop, called an *RS* flip flop (reset-set), is shown in Fig. 9-17. A MOS implementation that is area-effective uses two cross-connected two-input NOR gates. The accompanying truth table indicates that when both *R* and *S* (reset-set) inputs are 0, the *Q* and $\overline{Q}$ outputs will remain unchanged from its previous states. When *R* alone is a 1, the flip flop will reset, i.e., *Q* output will be 0, and $\overline{Q}$ output will be 1. When *S* alone goes to 1, the flip flop will be set, i.e., *Q* output will be 1, and *Q* will be 0. When *both R* and *S* are 1s, both *Q* and $\overline{Q}$ will be 0. Strictly speaking, this is then an indeterminate state since Q is no longer the opposite of $\overline{Q}$, but the device does provide well-defined outputs.

Latches

The circuitry in Fig. 9-18 allows information present at the data lead to be transferred to the output and the Q output to follow the data lead D, as long

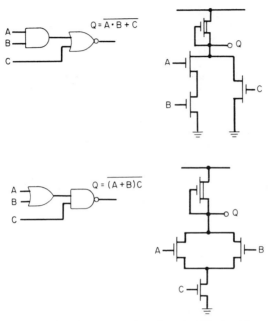

Figure 9-16 Implementation of two-level logic with a minimum number of devices.

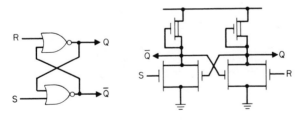

Inputs		Output
R	S	Q_n
0	0	Q_{n-1}
1	0	0
0	1	1
1	1	Indeter− minate

Figure 9-17 *RS* (reset-set) flip flop.

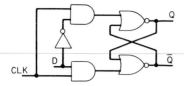

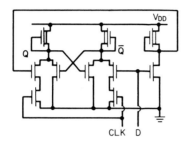

Figure 9-18 Latch circuitry where Q output follows data lead as long as CLK is high.

as clock CLK is a 1. But when CLK goes to a 0, then the information is latched in the flip flop until CLK goes high again. Figure 9-18 also illustrates the implementation of two-level logic using a single load device.

T Flip Flops

A *T* (toggle) flip flop changes state every time *T* goes to a 1. A *T* flip flop can be used as a binary countdown circuit (divide by 2) since *Q* will go through one period whenever *T* makes two transitions from 0 to 1. A logic representation of this function is given in Fig. 9-19. With *T* at a 0, all four AND gates are disabled and the two cross-coupled NOR gates are merely an *RS* flip flop with both set and reset inputs at a 0. When *T* goes to a 1, and only for a propagation delay until $\overline{T}$ goes to 0, *Q* and $\overline{Q}$ outputs are fed back in such a manner as to toggle the flip flop to an opposite state from which it started. Using both *T* and $\overline{T}$ (derived from *T*) to act simultaneously as enable signals results in an edge-triggered circuit which responds only on transition in *T*.

The circuit implementation is also shown in Fig. 9-19. When *T* is a 0 ($\overline{T}$ is a 1), the outputs are sampled and stored on the dynamic nodes with parasitic capacitances. On the rising edge of *T*, the flip flop is forced to change state. A propagation delay later, $\overline{T}$ goes to 0, preventing the flip flop from continuously toggling while *T* is high.

D Flip Flops and Shift Registers

A *D*-type flip flop is a circuit that takes data from an input lead at the beginning of a clock period, stores the information temporarily, and presents it to the output at the beginning of the next clock period. An implementation of this function is shown in Fig. 9-20a. At the beginning of the clock period, $\phi 1$ is high, and $\phi 2$ is low. Data are allowed to pass through to the first inverter on the left, which is decoupled from the second inverter. A master-slave relationship exists where the first inverter acts as master and receives data, while the second acts as slave and provides previous data to output Q. The second inverter holds the previous data by means of residual charge on its gate capacitance. When $\phi 1$ goes low and $\phi 2$ goes high at the second half of the clock cycle, the shift register is disconnected from its input and the two inverters are cross-coupled to transfer data effectively from master to slave. To ensure that the direction of transfer is from left to right, the feedback transistor is caused to have much lower gain by increasing its channel length L. To prevent direct feed-through of data, $\phi 1$ and $\phi 2$ should not overlap, i.e., should never be simultaneously high (see Fig. 9-20a).

D flip flops can be cascaded to form a shift register, allowing a stream of data to be stored in transit. With additional control circuitry, a recirculating memory results (Fig. 9-20b). When the lead marked STORE is a 1, data will recirculate through the shift register. When STORE is a 0, new data

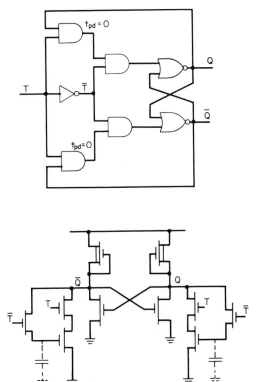

Figure 9-19 T (Toggle) flip flop.

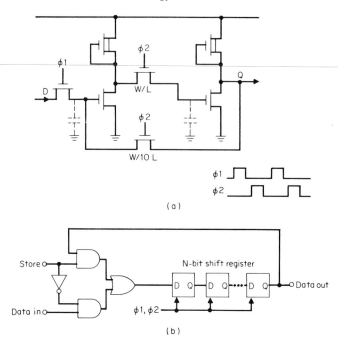

Figure 9-20 (*a*) *D* flip flop (*b*) **cascading to form a shift register for a circulating memory.**

are allowed to enter the shift register. The use of a shift register of course necessitates a counter to keep track of where the start of data is in the register.

Exclusive-OR Function

The Exclusive-OR function provides a 1 at the output only if either one, but not both, of its inputs is a 1. The truth table is shown in Fig. 9-21. It can be expressed as

$$(9\text{-}2) \quad A \oplus B = \overline{A} \cdot B + A \cdot \overline{B}$$
$$= \overline{(A + \overline{B})(\overline{A} + B)}$$

The last expression in Eq. (9-2) is a result of applying DeMorgan's theorem for logic equality and is the expression implemented in Fig. 9-21. In tracing through the truth table, one can see that the output will be a 1 only if the two inputs are dissimilar.

Schmitt Trigger

A Schmitt trigger provides a very rapid output transition when its slowly varying input crosses a particular trigger level. An essential part of such a circuit is to have a hysteresis in the transfer characteristics. Once the input crosses the trigger point in one direction and causes the output to change, then if it reverses direction, it has to pass that trigger point by a certain amount before the output will switch back. This prevents the output from toggling back and forth if noise happens to cause the input to jitter around the trigger point.

An implementation of the Schmitt trigger circuit in MOS is shown in Fig. 9-22a, together with the transfer curve showing the hysteresis. To understand how the circuit works, consider the case first where the input starts from 0 V and goes toward V_{DD}. Initially, all transistors are off and the output is a 1. When the input exceeds V_T, then M3 turns on forming an inverter with M4 as the load (Fig. 9-22b). Its output, node X, is the source of M2 and keeps M2 still off. As the input further increases, the voltage at node X decreases, tending to turn M2 on. When $V_{in} - V_x > V_T$, then positive feedback forces M2 to turn on rapidly, causing the output to snap to a 0.

When the input is going from a 1 to a 0, the circuit behaves very much like a two-input NAND gate (Fig. 9-22c) and the response is as shown.

A common use of the Schmitt trigger is in *power-on clear* circuitry as shown in Fig. 9-23. The slowly rising input voltage is used to generate a

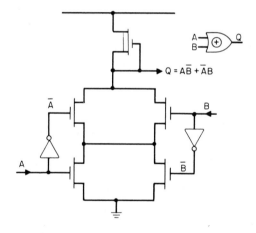

$$Q = A\bar{B} + \bar{A}B$$

Inputs		Output
A	B	Q
0	0	0
1	0	1
0	1	1
1	1	0

Figure 9-21 Exclusive-OR circuit.

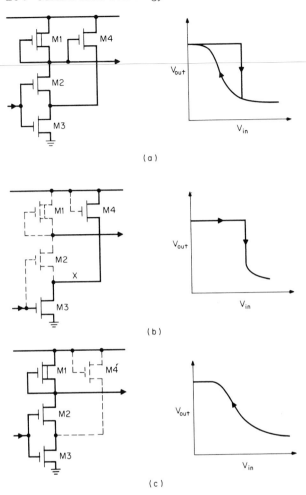

Figure 9-22 *(a)* **A Schmitt trigger circuit;** *(b)* **input going from 0 to 1;** *(c)* **input going from 1 to 0.**

reset signal that goes away a certain time period after power to the chip is applied. This reset signal is routed to all circuitry on the chip to ensure that everything starts from a known reset state.

Dynamic Logic

The MOS circuit implementations of logic functions described thus far have utilized the inverter as the basic building block. The inverter has a specified β ratio between the driver and load devices. This class of circuits is called *static ratioed logic*. A separate class of circuits can be designed where the load devices are gated by clock signals. Transmission of logic signals from one gate to the next is similarly gated. Since load and driver devices are never turned on at the same time, they both can be designed to be minimum size. This class of circuit is called *dynamic ratioless logic*. Dynamic logic refers to the fact that data between clocking periods are retained by capacitive charge storage.

Some advantages of dynamic ratioless logic are

1. Lower power dissipation, since for some configurations there is no direct discharge path from power supply lead to ground.

2. Smaller area due to extensive use of minimum area devices.

3. Inherent synchronous operation causes all signal propagation to be paced by the system clock. This minimizes timing problems.

An example[1] of dynamic ratioless logic is shown in Fig. 9-24. The circuit provides a 1-bit delay through two stages of inversion. The sequence of events is described in the following text and illustrated by the timing diagram of Fig. 9-24.

$\phi 1$ and $\phi 2$ are nonoverlapping clocks that never go high simultaneously. When $\phi 1$ is a 1, the input voltage (a 1) is coupled into capacitor $C1$ through pass transistor $Q1$, turning on $Q3$. Since both $Q2$ and $Q3$ are on, $C2$ will

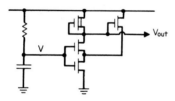

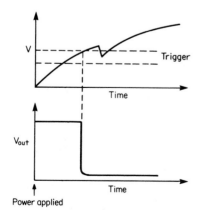

Figure 9-23 Power-on clear circuitry using Schmitt trigger.

(a) Logic diagram

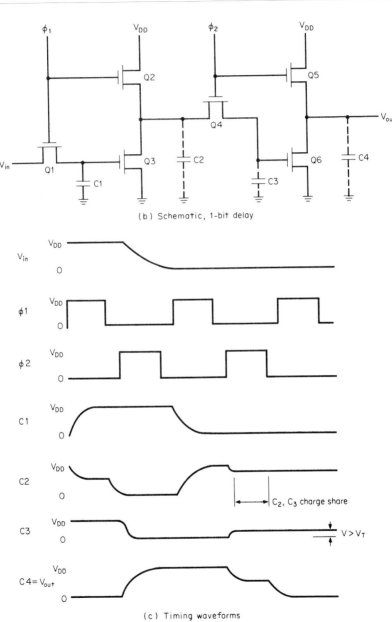

(b) Schematic, 1-bit delay

(c) Timing waveforms

Figure 9-24 Dynamic logic implementation of a one-bit delay.

be charged to a level which is voltage-divided between V_{DD} and ground. When $\phi 1$ returns to a 0, $Q1$ and $Q2$ are biased off, but the residual charge on $C2$ will keep $Q3$ on and discharge $C2$ to ground. On the second half of the clock cycle, when $\phi 2$ goes to a 1, $Q4$ and $Q5$ will be biased on. $C3$ will be discharged to ground through $Q4$ and $Q3$, keeping $Q6$ off and causing $C4$ to go to a 1. A 1 has thus been propagated from the input to output in one clock period.

If the input is a 0 when $\phi 1$ goes to a 1, then $C1$ is grounded, keeping $Q3$ off. Since $Q2$ is on, $C2$ will charge to approximately V_{DD}. When $\phi 1$ returns to 0, $Q2$ is biased off and $C2$ is floating. When $\phi 2$ goes to a 1, $Q4$ and $Q5$ are biased on, causing $C2$ to share charge with $C3$. Proper selection of ratio of capacitors $C2$ and $C3$ will provide enough voltage across $C3$ to bias $Q6$ on. $C4$ is now at a level which is voltage-divided between V_{DD} and ground, but when $\phi 2$ returns to ground, $C4$ discharges to a 0 through $Q6$. A 0 has thus been propagated from the input to output in one clock period.

In this example, there is a dc path between V_{DD} and ground, thus consuming dc power. Addition of other gating transistors can circumvent this disadvantage.[1]

9-5 MEMORY CIRCUITS

Memory circuits form an essential part of any digital system. The use of memory circuits in such systems has continued to accelerate, sustained to a large extent by the ability of the technology to provide continually denser memory chips. High density translates into lower cost and higher speed. The volume in standard memory products outpaces by far any other types of MOS circuits; but memory circuits are also found imbedded within logic chips such as microprocessors.

RAM (Random Access Memory)

A serial memory was described in Sec. 9-4, "Shift Registers," where the access time to data in different memory locations will vary. In a RAM, each bit of information is accessible individually by applying the appropriate address (combination of 1s and 0s) to the address leads. The access time, the delay from the time the address leads change to the time valid data appear on the output, is an important specification parameter and is practically the same for all bits. RAMs can be separated into static RAMs and dynamic RAMs.

STATIC RAM

A static RAM is one in which the data are retained as long as power is applied, though no clocks are running. Figure 9-25 shows the most common implementation in which six transistors are needed. When this particular

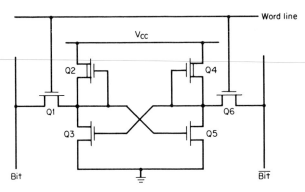

Figure 9-25   **Six-transistor static RAM cell.**

cell is selected, the word line (usually a continuous polysilicon stripe) turns on the two pass transistors $Q1$ and $Q6$, allowing BIT and $\overline{\text{BIT}}$ lines to remain at a precharged high state or be discharged to a low state by the flip flop. Differential sensing of the state of the flip flop is then possible. In writing data into the selected cell, BIT and $\overline{\text{BIT}}$ are forced high or low by some other WRITE circuitry. The side that goes to a 0 is the one most effective in causing the flip flop to change state.

How a complete static memory works can perhaps be best explained with the hypothetical 16-bit RAM[5] illustrated in Fig. 9-26. The cells are physically arranged in an *XY* format, usually square. Half of the address bits (A_0 and A_1) are decoded to select one out of four word lines (rows). The other half of the address bits (A_2 and A_3) are used to select one out of the four pairs of bit lines (columns). During a WRITE operation, after the proper row and column are selected, the data-in lead DI forces bit sense columns BSC and $\overline{\text{BSC}}$ to the appropriate levels, forcing the flip flop to the desired state. During a READ operation, after the appropriate cell is selected, the signals on BSC and $\overline{\text{BSC}}$ are applied to the sense amplifier. In Fig. 9-26, the sense amplifier is a cascade of three differential amplifier stages that provides gain to speed up the voltage transition, provides dc level shifting, and also drives the output (DO) buffer.

Other leads needed for a fully functional memory circuit are the write-enable WE, which selects between WRITE and READ modes, and the chip-enable $\overline{\text{CE}}$, which when at a 1, disables the sense amplifiers and output.

DYNAMIC RAMS

A dynamic RAM stores its data as charge on a capacitor. Even if power is not interrupted, charge can leak off due to various leakage current paths, and the data need to be refreshed every few milliseconds. Figure 9-27*a* shows the simplest dynamic memory cell to consist of a single transistor in

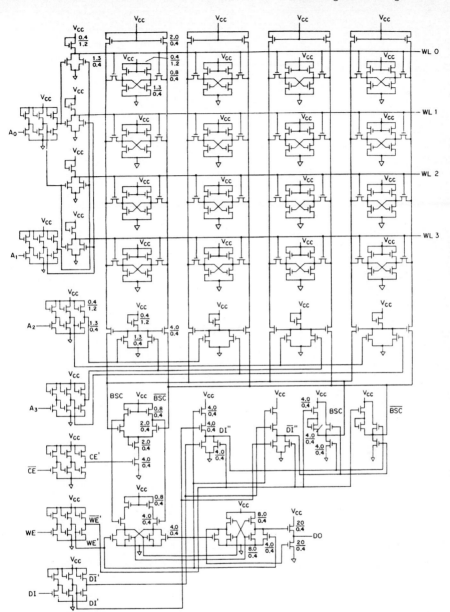

Figure 9-26 Hypothetical 16-bit RAM circuitry. (*After Ref. 5.*)

series with a data storage capacitor.[6] Accessing an individual cell again requires the selection of a word line (row) which turns on the transfer gate transistor, and the selection of a bit line (column) on which the data are placed. During a read cycle, the storage capacitance shares charge with the bit line capacitance. The ratio of the two capacitances has to be selected

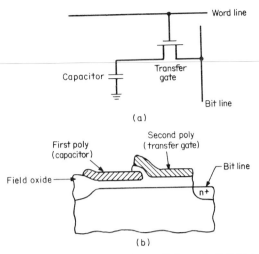

Figure 9-27 A one-transistor cell dynamic RAM (*a*) Circuit; (*b*) implementation.

properly to ensure that the resultant voltage can be detected by the sense amplifier. Unlike the static memory, reading in this case is normally destructive and therefore must be followed by a write cycle. However, present circuit design allows the act of latching the data onto the sense amplifier to rewrite the cell automatically, so no subsequent write cycle is necessary. Systematically reading each cell in sequence and automatically writing the information back into the cell would constitute a refresh cycle. Certain newer designs allow this refresh cycle to be carried out automatically by the chip's internal logic.

One way to lay out a very dense memory cell, shown in Fig. 9-27*b*, is by using two layers of overlapping polysilicon. The second (top) layer forms the gate of the transfer transistor, and the data are stored as an inversion charge in the first-layer poly MOS capacitor. This is very similar to the transfer and storage of charge in a charge-coupled device (CCD). The first-layer poly is tied to some dc voltage.

Since the data in a dynamic memory are stored in such a small capacitance of less than 100 fF, they are susceptible to upset by alpha particles.[7] Alpha particles are emitted from natural radioactivity in the IC package, such as from the glass used to seal the package. These particles create electron-hole pairs when they penetrate the depletion layer of the MOS capacitor. The electrons are collected at the surface. Electrons from electron-hole pairs generated outside of the depletion region can also diffuse toward the surface and get collected. The result is not a hard failure but rather a three to five orders of magnitude increase in the error rate, giving rise to "soft errors." Solutions found to be effective include coating the memory chip with a thick layer of inorganic polyimide film, or better yet, designing the part to have higher storage capacitance such as using nitride as dielectric.

ROM (Read-Only Memory)

A read-only memory is one wherein the information contained in the memory cannot be readily changed. Otherwise, it acts like a regular memory in the sense that when an address is applied, one bit of information appears at the output. Unlike a RAM, the ROM will retain its memory even if the power is turned off. It is the natural choice for microprocessor systems to store microinstructions for program execution (such as the BASIC interpreter) so that those programs will be available the moment power is applied.

A ROM is shown in Fig. 9-28. Part of the address word is used to do a row-select. The presence or absence of a transistor constitutes the ROM coding, and causes data to appear on the column lines. The latter are in effect the outputs of multi-input NOR gates. The rest of the address word is used for column select to pick the proper column for the output. Figure 9-28 also illustrates the fact that a ROM can often have more than one output. In that case, an input address produces an output "word," i.e., several bits of data.

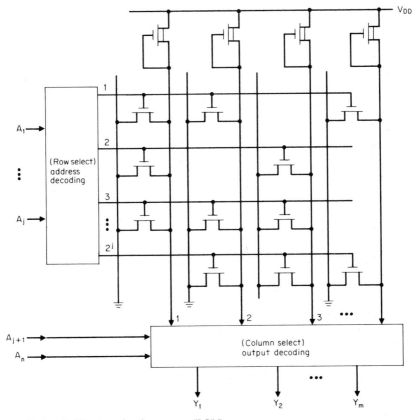

Figure 9-28 A read-only memory (ROM).

The presence or absence of a transistor can be implemented by programming the depletion, diffusion, contact, or metal mask. The first two result in the densest layout, but the last two produce the shortest turnaround time from programming the mask to parts on hand.

Another application for ROMs is for code converters such as from four-bit hexadecimal code to seven-segment LED (light-emitting diode) display. ROMs are also used as look-up tables. For example, to find $f(x^2)$, a two-bit binary input results in a four-bit output. In general, when there are n bits of address and m bits of output word, then the total storage capacity is $m \cdot 2^n$.

While a ROM is best suited for accessing a bank of systematically arranged data, its orderly implementation also makes it suitable for random logic. One property of combinational logic is that a logic function can be reduced to an expression of the general form

(9-3) $\quad F = A \cdot B \cdot C + \cdots + D \cdot E \cdot F + \cdots$

This is the sum of product form, and each product term is called a *minterm*. This implies that the function can be implemented with two levels of logic, the first being an AND level and the second an OR level. This is the key concept behind *programmable logic arrays*[8,9] (PLA). A PLA example is shown in Fig. 9-29. The first ROM array on the top is the AND plane, and the ROM array on the bottom is the OR plane. Both planes are simply arrays of large, multiinput NOR gates. The top plane results in an AND function by virtue of applying DeMorgan's theorem which says that inverting all inputs will change an OR function to an AND.

(9-4) $\quad \overline{A} + \overline{B} = A \cdot B$

Note from Fig. 9-29 the orderly implementation of the functions. A further advantage of PLA results if certain address words are not used, or if the output is the same for a different address, then certain rows and columns in the array can be deleted. In that respect, a PLA can be considered a subset of ROM.

PROM (Programmable Read-Only Memory)

Standard ROMs are programmed during fabrication by the information encoded in a certain mask level. By the nature of its encoding, the programming is permanent. There is a class of ROM, usually in bipolar technology, that allows the user to program the code by blowing fusible links (nichrome or polysilicon). Those ROMs, also called PROMs, are also unalterable once programmed, but are not the subject of this section. The discussion centers, rather, on a class of ROMs that allows the end user to erase the previous information and then program in new data. This type of device is implemented in MOS technology. The need for a separate ERASE cycle makes

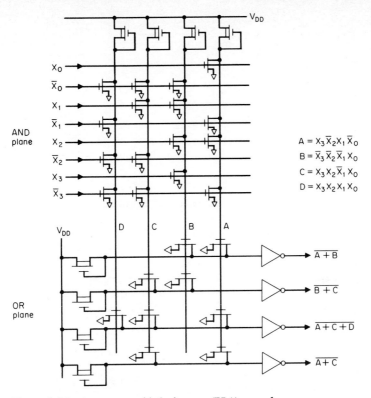

$A = X_3 \overline{X}_2 X_1 \overline{X}_0$
$B = \overline{X}_3 \overline{X}_2 \overline{X}_1 X_0$
$C = X_3 X_2 \overline{X}_1 X_0$
$D = X_3 X_2 X_1 X_0$

Figure 9-29 **A programmable logic array (PLA) example.**

such a part best suited for applications where data are not changed too frequently; otherwise, RAMs would be more appropriate. On the other hand, PROMs do have the advantage of nonvolatility, i.e., they retain their information even when power is not applied.

Two major types of PROMs are the electrically programmable ROM (EPROM) and the electrically erasable ROM (EEPROM).

EPROM

An electrically programmable ROM is programmed by having hot electrons injected into a floating gate and causing a substantial shift in the threshold. This is the floating gate avalanche-injection MOS (FAMOS) device as shown in Fig. 9-30a. Under high gate *and* high drain voltages, electrons gain sufficient energy to jump the silicon–silicon dioxide energy barrier, penetrating the oxide and flowing to the floating gate, which is completely surrounded by oxide.[10] The injected electrons cause a 5- to 10-V increase in the threshold of the device, changing it from an ON to an OFF state when

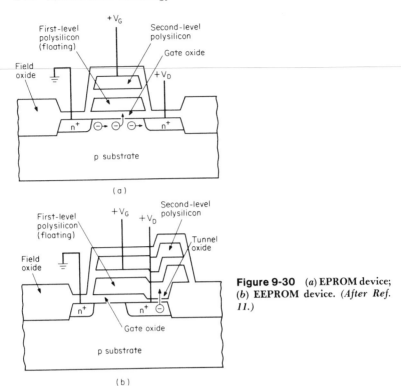

Figure 9-30 (*a*) EPROM device; (*b*) EEPROM device. (*After Ref. 11.*)

a 5-V READ voltage is applied to the gate. Once programmed, a cell can hold its charge for 10 to 100 years.

The READ-WRITE cycle can be explained with the aid of Fig. 9-31 which shows four single-transistor cells. Writing (injecting electrons) into the upper left cell is accomplished by pulsing the row-select line to a high programming voltage, $V_{pp} = 20$ V for 10 ms, while raising its corresponding drain column line to 18 V. Nonselected rows and columns are at 0 V. Reading consists of applying 5 V to both selected row and column. A written cell produces no current flow, causing the column line to go to a 1.

Erasing an EPROM memory requires taking the IC out of its socket and exposing it to intense UV light for 20 min. A transparent quartz window is built into the top of the IC package for this purpose. Programming (writing) is also performed off the equipment on a separate programming instrument. Because of this inconvenience and delay in changing the program, EPROMs were originally envisioned as a development tool for designers who change programs frequently while prototyping and debugging a system. Now, EPROMs are often used in production equipment where many versions or updates are required. In such a case, the production volume for each version does not justify the time and expense for a mask programmed ROM.

An EPROM memory can be erased and written over several hundred times.

EEPROM

An electrically erasable PROM allows the erasure of data by electrical means, opening up the flexibility of in-circuit reprogramming.[11] EEPROM cells require the application of a gate field equal or greater than 10^7 V/cm on a very thin (less than 200 Å) gate oxide, such that Fowler-Nordheim tunneling of electrons into the oxide will take place. The tunneled charge shifts the threshold voltage by 5 to 10 V. Since tunneling is bidirectional, to erase, the gate field is simply reversed.

One form of an EEPROM is an MNOS device where a nitride layer is placed on top of the gate oxide and the tunneled electrons are retained at the oxide-nitride interface. This type of device is more commonly referred to as electrically alterable ROM (EAROM). Another implementation of EEPROM is to use a floating gate, as shown in Fig. 9-30*b*. This device

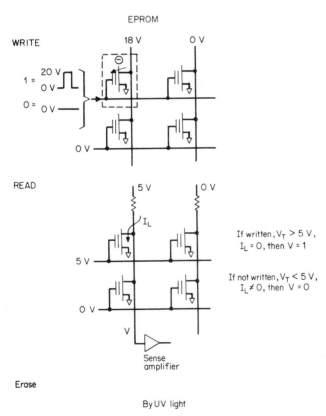

Figure 9-31 **WRITE, READ, and ERASE** cycle for an **EPROM.**

resembles the FAMOS structure except for a small tunnel oxide region between the gate and the drain. Confining the tunnel oxide region to a small area helps maintain the overall gate oxide integrity.

The WRITE-READ-ERASE cycle can be explained with the aid of the four cells drawn in Fig. 9-32. Two transistors are required per cell. During WRITE the upper left cell is selected by applying 25 V to the row-select and grounding the column line. Electrons will tunnel to the floating gate if a 10-ms 25-V pulse is applied to the V_{pp} line. Other cells in the same row will not be programmed if their column line is raised to 25 V, keeping the drain-to-gate potential of the EEPROM cell at zero. Nonselected rows keep their programming lines at zero. During READ, 5 V is applied to both row-select and column lines. A programmed cell will have a threshold greater than 5 V, producing zero current and a 1 at the column line. Conversely, a nonprogrammed cell will conduct, driving the column line to 0. During ERASE, applying 25 V to both the selected column and row and keeping the programming line at 0 V will allow electrons to tunnel back from the floating gate to the drain.

It is not necessary to work with simply one bit at a time. Most EEPROM memories allow "byte-wide" ERASE and WRITE by manipulating more than one column at a time. EPROM can also be written by bytes, but its ERASE cycle is of course block erase.

With EEPROMs, the current flow directions for cell writing and erasing are arbitrary. The direction adopted here is made consistent with that for EPROM, that is, writing is forcing electrons toward the floating gate. The reverse convention is to call ERASE the act of forcing electrons onto *all* floating gates and selectively write a 0 by discharging the cell.

9-6   OTHER CIRCUIT TECHNIQUES

Output Buffers

A simple inverter to drive output loads (usually capacitive) will result in uneven rise- and falltimes. Increasing the load transistor size to improve the risetime will inordinately increase the dc power consumption and area of the inverter. One solution is to use a push-pull output driver such as shown in Fig. 9-33a, which also inverts the outgoing signal. When the output pull-down device is being turned on, the pull-up device is simultaneously being turned off. Conversely, when the pull-up device is being turned on, the pull-down device is shutting off. Such a ratioless configuration allows individual tailoring of device sizes for symmetrical waveforms.

Certain push-pull buffers also allow a *tristate* output, where the third state is a high-impedance path to either power supply rail. Figure 9-33b shows that when the tristate lead TS goes high, both push-pull devices are shut off, allowing the output lead to float. An output lead that is in tristate can then be controlled by other chips with minimum interactions between

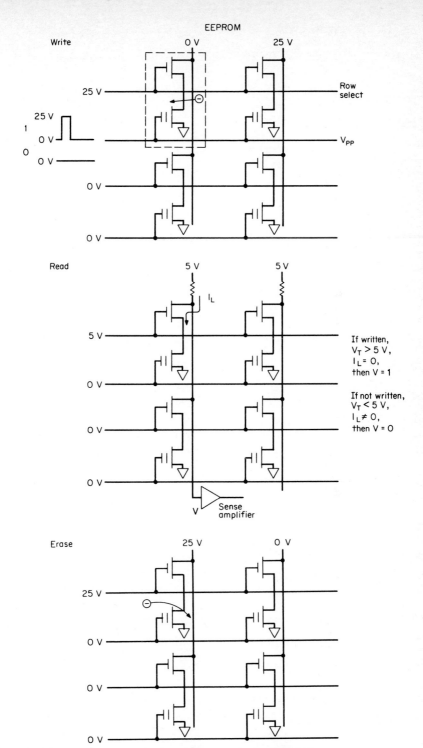

Figure 9-32 WRITE, READ, and ERASE cycle for EEPROM.

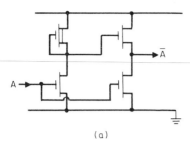

(a)

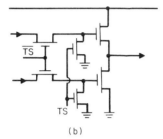

(b)

Figure 9-33 (*a*) Inverting version of push-pull output buffer; (*b*) addition of circuitry for tristate.

the ICs. Circuit speed will improve with the use of depletion device for pull-up instead of enhancement as shown; however, the tristate capability cannot be added.

A noninverting configuration results by reversing the connections to the pull-up and pull-down devices. In both inverting and noninverting configurations, a large current spike flows for the short period when both output devices are on.

Bootstrap Circuits

It has been shown that by keeping the load device of an inverter always turned on, the output risetime is reduced. The risetime can be speeded up even further by providing an overdrive to the load device's gate. Such circuits are called *bootstrap* circuits since certain node voltages are self-coupled to higher than normal levels, even though only momentarily.

One such circuit is shown in Fig. 9-34. When transistor $Q2$ is turned on, a voltage of $V_{DD} - V_T$ is developed across capacitor $C1$. When input voltage shuts off $Q2$, the output voltage starts to rise toward V_{DD}. The stored charge in $C1$ is of such a polarity as to push the voltage at node A higher, even beyond V_{DD}. This provides the overdrive for $Q3$. The increase in speed is obtained with virtually no extra power expended. Figure 9-34 shows that after a few milliseconds, leakage current of junctions connected to the capacitor will drain off the stored charge, but the speedup has already been accomplished. A long channel length depletion device ($Q4$) is connected at the output to provide the final pull-up to V_{DD} level.

The higher voltage appearing on node A necessitates careful consideration to assure that junction breakdown or transistor punch-through is not encountered. Also, any parasitic capacitance associated with node A would reduce the bootstrapping action and should be minimized.

Substrate Bias Generator

Providing a back-bias to the substrate with respect to the source (system ground) provides numerous circuit advantages. Device threshold voltages are increased, but field threshold voltages are increased by an even larger amount. The drain junction capacitance is lowered, providing improvements in speed. Although the use of ion implantation has made back-bias no longer necessary for threshold adjust, the reduction in drain capacitance still makes it a desirable mode of operation. The availability of an on-chip substrate bias generator makes the continued use of back-bias viable.

A substrate bias generator is shown in Fig. 9-35. A square wave oscillator provides a normal 0- to 5-V signal at the input. On the positive rising edge, transistor $Q1$ (acting like a diode) turns on and clamps the voltage at

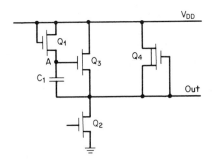

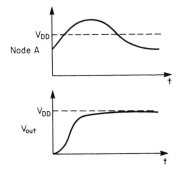

Figure 9-34 A bootstrap circuit to speed up output risetime.

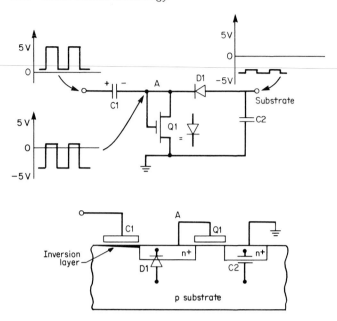

Figure 9-35 A substrate bias generator.

node *A* to one threshold drop above ground. The rest of the 5-V signal is developed across the capacitor *C*1 with the direction as shown. On the negative going edge, *Q*1 is turned off and node *A* is driven negative, pulling the substrate negative with it. Diode *D*1 and capacitor *C*2 keep the substrate at negative potential even when node *A* swings positive again. The back-bias generator only needs to supply the junction leakage and other sources of leakage current.

The implementation of the circuit, shown also in Fig. 9-35, allows merging of several components of the circuit. For better performance, diode *D*1 can be replaced with a transistor connected as a diode, similar to *Q*1. In such a case, a separate bonding connection needs to be made from its drain to the substrate.

REFERENCES

1. W. M. Penney and L. Lau (eds.), *MOS Integrated Circuits,* Chaps. 4 and 5, Van Nostrand Reinhold, New York, 1972.

2. R. H. Crawford, *MOSFET in Circuit Design,* Chap. 5, McGraw-Hill, New York, 1967.

3. A. B. Glaser and G. E. Subak-Sharpe, *Integrated Circuit Engineering: Design, Fabrication and Applications,* Chap. 14, Addison-Wesley, Reading, Mass., 1977.

4. D. J. Hamilton and W. G. Howard, *Basic Integrated Circuit Engineering,* Chap. 14, McGraw-Hill, New York, 1975.

5. Courtesy of L. T. Lin, Motorola, Inc.

6. V. L. Rideout, "One-device cells for dynamic random-access memories: A tutorial," *IEEE Trans. Electr. Dev.,* vol. ED-26, pp. 839–852, June 1979.

7. T. C. May and M. H. Woods, "Alpha-particle-induced soft errors in dynamic memories," *IEEE Trans. Electr. Dev.,* vol. ED-26, pp. 2–9, January 1979.

8. W. N. Carr and J. P. Mize, *MOS/LSI Design and Applications,* Chap. 8, McGraw-Hill, New York, 1972.

9. C. Mead and L. Conway, *Introduction to VLSI Systems,* Chap. 3, Addison-Wesley, Reading, Mass., 1980.

10. D. Frohman-Bentchkowsky, "A fully decoded 2048-bit electrically programmable FAMOS read-only memory," *IEEE J. Solid-State Circ.,* vol. SC-6, pp. 301–306, October 1971.

11. W. S. Johnson et al., "16-K EEPROM relies on tunneling for byte-erasable program storage," *Electronics,* pp. 113–117, Feb. 28, 1980.

PROBLEMS

1. A saturated load device inverter whose transfer characteristics are given by Fig. 9-7 is driven by an identical inverter whose driver device is OFF. If the W/L ratio of the driver device is 10, what would be the W/L ratio of the load device in order to have a 0 output level smaller than 2 V? $V_{DD} = 12$ V.

2. The inverter described by Fig. 9-10 is driven by an identical inverter. Graphically determine its noise margin when its input is a 1 (NM^1) and when its input is a 0 (NM^0). The $\beta_R = 10$.

3. From a speed standpoint, which power supply and temperature limits are the worst case? Why?

4. Based on the computer simulations represented in Fig. 9-15, what are the propagation delays when output is rising (t_{pd}^+) and when output is falling (t_{pd}^-) for the inverter whose load device W/L ratio is 1.0? The load capacitance is 1.0 pF. The $\beta_R = 10$.

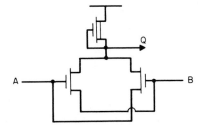

Figure 9-36 Prob. 6.

5. Generate a timing diagram for all input, output, and critical internal nodes for the toggle flip flop of Fig. 9-19 to illustrate its function.

6. Construct a truth table for the circuitry in Fig. 9-36 to determine its function. Considering that the inputs now have to sink dc current flow, what might be one source of problem? Consider in particular if there is large resistance in series with inputs A or B.

10

Analog MOS Design

10-1 CONSIDERATIONS IN ANALOG MOS CIRCUITS

MOS transistors saw widespread application first in digital circuitry, which was more tolerant of variations in device gains and device thresholds. Since the late 1970s, interest in using MOS for analog circuits has been increasing rapidly. As circuits have advanced to higher levels of integration, there has been a need to interface directly with external world analog signals. MOS processing has improved to the appropriate level of consistency and uniformity such that analog design with MOS becomes practical.

MOS circuits excel over bipolars in density in analog circuits because they are self-isolating. A MOS operational amplifier (op-amp) takes 100 to 200 mil^2, which is one-third to one-half of the die area of equivalent bipolar versions and consumes 1 to 10 mW. With such small size, and low power, using 20 to 30 op-amps in a single chip is not at all uncommon. MOS circuits also have the unique ability to store charge on a node for periods of several milliseconds and use a high-input impedance MOS transistor to sense this charge nondestructively.

Bipolar transistors still have higher transconductance, higher output drive capability, and lower noise. In terms of absolute performance, bipolars will therefore prevail. For example, there are only a few single MOS op-amps packaged for sale as components. However, the ability to realize digital and analog circuits on the same large-scale integration LSI technology has pushed MOS technology into the realm of linear design.

Transconductance g_m

Transconductance is the increase in drain current due to an increase in gate voltage. For MOS devices, transconductance is highest in saturation, thus

223

most of them are operated in that region for analog circuits. Since

(10-1) $\quad I_D = \tfrac{1}{2}\beta(V_G - V_T)^2$

therefore

(10-2) $\quad g_m = \dfrac{\partial I_D}{\partial V_G}\bigg|_{V_{D,\text{constant}}} = \beta(V_G - V_T)$

$\qquad\quad = \sqrt{2 I_D \beta}$

Equation (10-2) states that the transconductance of a MOS transistor increases only as the square root of the drain current. Bipolar transistors, instead, have transconductance that is proportional to the emitter current I_E. Since bipolars are usually operated in the milliampere range while MOS devices are in the microampere range, bipolar transistors have inherently much higher gain. As a result, MOS amplifiers need more stages to achieve the same amount of gain. Or, for the same output current drive, a larger gate-to-source voltage is needed.

Amplifier Gain

A basic amplifying gain stage is an inverting amplifier consisting of the active device and a load, as shown in Fig. 10-1. Also shown is its equivalent circuit for ac analysis. The gain of the stage is given by

(10-3) $\quad A_V = \dfrac{V_{\text{out}}}{V_{\text{in}}} = -g_m(R_L || r_o)$

The output impedance r_o is an important element of, and often dominates, the total load resistance. In the ideal case, the output impedance of a MOS transistor is infinite in saturation. In practice, it has a finite value. A first-order description of this finite output impedance, which is caused by the

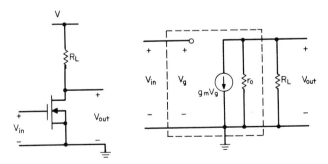

Figure 10-1 An inverting amplifier with its equivalent circuit.

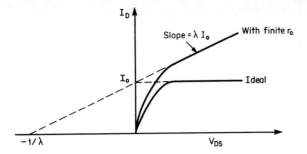

Figure 10-2 Finite output impedance due to channel-short-ening effect.

channel-shortening effect, is given by modifying the saturation drain current equation:

$$(10\text{-}4) \quad I_D = \frac{\beta}{2}(V_G - V_T)^2(1 + \lambda V_{DS})$$

$$= I_o(1 + \lambda V_{DS})$$

A graphic description of Eq. (10-4) is given by Fig. 10-2. The slope of the curve in saturation is given by λI_o. The intercept with the negative X axis is $-1/\lambda$, analogous to the Early voltage for bipolars. The ideal saturation current I_o is the intercept with the Y axis. The output impedance is the inverse of the slope of the curve:

$$(10\text{-}5) \quad r_o = \frac{\partial V_{DS}}{\partial I_D} = \frac{1}{\lambda I_o}$$

Lambda (λ) decreases with increasing channel length and substrate doping. Its precise value is best measured from actual devices, but for a gate length of 10 μm, a typical value[1] is 0.03 V^{-1}. Problem 1 provides a way to calculate for λ.

Device Capacitance

Design of analog MOS circuits depends heavily on simulation programs (see Chap. 12) which contain device models with proper analog character-istics. One important characteristic is the device capacitance.

In the off state, capacitance between the gate and the source or drain consists of overlap capacitance between the gate and source and drain junc-tions, while the gate-to-substrate capacitance is the oxide capacitance rep-resented by the gate electrode area (see Fig. 10-3). In the triode state, the formation of the channel shields the gate from the substrate. The oxide

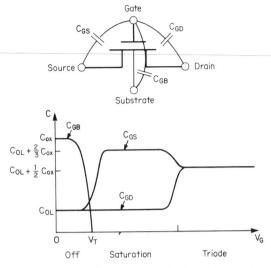

Figure 10-3 Device capacitance in the various modes of operation of the transistor.

capacitance is now equally allocated to both source and drain. In the saturation state, the channel pinches off and pulls back from the drain. The gate-to-drain capacitance reverts to the overlap value, while two-thirds of the gate oxide capacitance is assigned to between the gate and source.

A graphical representation of all three conditions is given in Fig. 10-3 where all the capacitance components are plotted as a function of gate voltage. As gate voltage is increased from zero, the transistor passes from the off state, through saturation, and to the triode state. The partitioning of the gate oxide capacitance between the source and drain in triode and saturation is somewhat arbitrary but agrees well with actual device measurements.

Noise

MOS transistors tend to be noisier than bipolars, and thus noise performance is often a major concern in MOS circuits, particularly at low (audio) frequencies. One component is the flicker noise, or $1/f$ noise.[2,3] This arises from the charging and discharging of interface traps (surface states) of density D_{it} at the oxide-semiconductor interface (see Sec. 4-7, "D_{it} Interface Traps"), and of traps in the oxide situated very near the silicon surface.

When referred to the input of the transistor, the value of the equivalent noise source has a mean-squared voltage of

$$(10\text{-}6) \qquad \overline{V}_{eq}^2 = \frac{K}{C_o WL} \frac{1}{f} \Delta f$$

where C_o = oxide capacitance per unit area
$W \times L$ = area of gate
f = measurement frequency
Δf = band width of the system

K is a constant with a typical value of $3 \times 10^{-24} \text{ V}^2 \cdot \text{F/Hz}$. This yields a noise density of 100 nV/$\sqrt{\text{Hz}}$ at 1 kHz for a 100 μm $\times$ 10 μm NMOS device with 1200-Å gate. Noise spectral density is shown in Fig. 10-4. The $1/f$ function is evident at low frequencies. At frequencies higher than around 10 kHz, a constant thermal noise, that arises because of the resistance of the channel, dominates. When referred to the input, its equivalent input noise, voltage, though highly variable from process to process, is usually given by

$$(10\text{-}7) \quad \overline{V}_{eq}^2 = 4kT \left(\frac{2}{3} \frac{1}{g_m} \right) \Delta f$$

This component is dependent on temperature and the two factors affecting transconductance, namely bias current and W/L ratio.

10-2 ANALOG BUILDING BLOCKS

Voltage Gain Blocks

A voltage gain block consists of the simple amplifier introduced in Sec. 10-1, "Amplifier Gain." There are three versions of such an amplifier, depending on the type of load device used.

SATURATION LOAD[4]

The inverter configuration containing a saturated load device, which is used in digital circuits, can also amplify ac signals. Figure 10-5a shows the circuit with the transfer characteristics. The voltage gain can be derived by equating the driver and load currents and redefining the gate-to-source voltage of

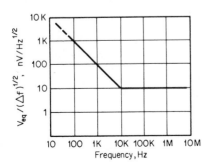

Figure 10-4 Noise spectral density for a typical NMOS device.

the load device. That is,

$$(10\text{-}8) \quad I_L = I_D \quad \text{and} \quad V_{\text{GSL}} = V_{DD} - V_{\text{out}}$$

The gain, neglecting change in threshold due to back-gate bias, is

$$(10\text{-}9) \quad A_V = \frac{\partial V_{\text{out}}}{\partial V_{\text{in}}} = -\sqrt{\frac{(W/L)_{\text{driver}}}{(W/L)_{\text{load}}}} = -\sqrt{\beta_R}$$

The gain is caused by geometrical factors (W/L) only, and the temperature and device gain factors β_o have canceled out. The nonlinearity of the load has also canceled out, producing a large linear region as shown. However, the gain is rather low, going as the square root of the beta ratio. A voltage gain of no more than 5 to 10 is achievable before the area required becomes inordinately large.

DEPLETION LOAD

A depletion-load device, connected in the manner shown in Fig. 10-5b, ideally acts like a current source. Its output impedance, limited only by the channel-shortening effect, is extremely high, resulting in very high amplifier gain as indicated by Eq. (10-3). However, the back-gate bias effect reduces the gain substantially. A small-signal analysis[5] shows the voltage gain to be

$$(10\text{-}10) \quad A_V = -\frac{2}{\gamma} \sqrt{\beta_R} \sqrt{V_O + V_{BB} + 2\phi_f}$$

where β_R = beta ratio
  V_O = output dc voltage
  V_{BB} = substrate bias with respect to ground
  γ = body effect constant = $\sqrt{2\epsilon_s\epsilon_o q N_A}/C_o$

For the same beta ratio, using a depletion load gives four to six times higher gain than a saturated load device. Voltage gains of 25 to 50 are typical. Equation (10-10) indicates that voltage gain is improved by operating with a substrate bias V_{BB} and by biasing the output voltage away from ground. To obtain equal positive and negative swings, the output is normally biased midway between V_{DD} and ground.

The transfer characteristics of a depletion-load amplifier are given also in Fig. 10-5b. Because a change in output voltage also changes the back-gate bias, the transfer characteristics are slightly nonlinear.

COMPLEMENTARY LOAD

The problem of back-gate bias associated with the use of depletion loads is solved by using a complementary-load device whose source is tied to its sub-

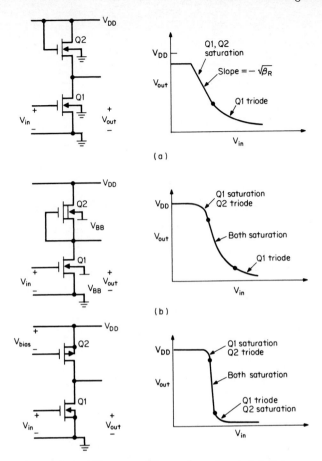

Figure 10-5 Different amplifier configurations and their transfer characteristics: (*a*) with saturated load, (*b*) with depletion load, (*c*) with complementary load.

strate. Figure 10-5*c* shows the use of a *p*-channel device as the load for an *n*-channel driver. The *p*-channel device is operated in the saturation mode with a constant gate-to-source bias, functioning as a current source with a high output impedance.

The transfer characteristics are also shown in the figure. In the linear region, both transistors are in saturation. When either one of them is in triode, the amplifier is highly nonlinear. The gain in the linear region is the transconductance of the driver device $Q1$, multiplied by the effective output resistance of $Q1$ and $Q2$ in parallel. These quantities are given by

$$(10\text{-}11) \quad g_{m1} = \sqrt{2I_D\beta_1}$$

and

(10-12) $r_{o1} = \dfrac{1}{\lambda_1 I_D}$ and $r_{o2} = \dfrac{1}{\lambda_2 I_D}$

Thus

(10-13) $A_V = \dfrac{1}{\sqrt{I_D}} \dfrac{1}{\lambda_1 + \lambda_2} \sqrt{2\beta_1}$

Although the transconductance increases with drain current, it increases only as the square root of I_D, while the output resistance is inversely dependent on the first power of I_D. The net result is that higher voltage gain is achieved with *decreasing* drain current. Voltage gain will continue increasing until the drain current is reduced such that the device enters the subthreshold region. The transconductance then becomes directly proportional to current and the gain reaches a maximum constant value. Voltage gains of 200 to 1000 V are typical for a few tenths of a microampere drain current. Excessive noise often precludes operation with extremely small drain current.

The load lines for all three types of loads are superimposed on the output characteristics of the driver device in Fig. 10-6. Note the difference in output resistance, represented by the slope of the load lines. The figure also shows that normally the output voltage is biased at $V_{DD}/2$.

Current Sources

Current sources are used extensively in MOS analog circuits, both as biasing elements and as active loads to obtain high ac voltage gain. Just as for bipolars, MOS current sources are current mirrors constructed by passing

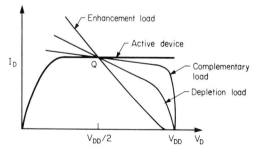

Figure 10-6 Load lines for saturated, depletion, and complementary loads, superimposed on the characteristic curve of a driver device. Quiescent point Q is normally at $V_{DD}/2$.

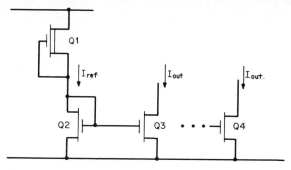

Figure 10-7 MOS current source.

a reference current through a diode-connected (gate tied to drain) transistor. The voltage developed across it is applied to the gate and the source of a second transistor which provides the output current (Fig. 10-7). To increase the output impedance of the current source, the channel length of transistor $Q3$ should be large. The W/L ratio of $Q3$ can be directly scaled up or down to make I_{out} either smaller or larger than I_{ref}. It is also possible, and in fact quite common, to have several current source outputs by placing other devices in parallel with $Q3$ as the figure shows. All output currents will scale up or down together if I_{ref} changes, such as from power supply or temperature fluctuations. The use of current sources for biasing (to the extent that they are ideal current sources) makes the circuit performances insensitive to such fluctuations.

Output Buffers

In order to drive a resistive or large capacitive load, an output buffer or driver is usually called for. In order not to degrade the stability of the amplifier by additional phase shift, a unity-gain voltage follower configuration is often used. A basic output buffer is shown in Fig. 10-8a. It is a source follower with a constant current source as load device. Such a circuit is a class A amplifier, i.e., there is dc flow through both devices on each half of the ac swing. This buffer is highly linear, but suffers from the limitation that the largest current it can sink is equal to the bias current. As with all class A amplifiers, this buffer dissipates high quiescent power.

A variation on Fig. 10-8a is to replace the driver MOS transistor $Q1$ with a common collector bipolar *npn* transistor. This is possible with CMOS processes that have *p*-tubs into an *n*-type substrate. A parasitic transistor is available whose collector is the *n* substrate (always connected to V_{DD}), whose *p*-type base is the *p*-tub and the emitter is the *n*-channel source or drain diffusion inside the well. The *npn* transistor is capable of current gains of about 500 but has some slight propensity for initiating

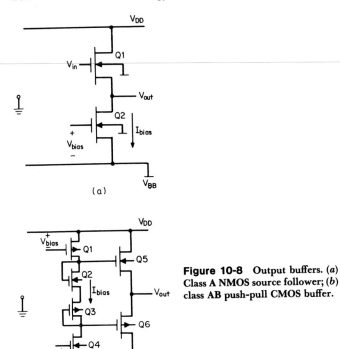

Figure 10-8 Output buffers. (*a*) Class A NMOS source follower; (*b*) class AB push-pull CMOS buffer.

CMOS latch-up (see Chap. 11) with its forward-biased emitter-base junction.

A class AB amplifier provides high ac drive while drawing low quiescent power. An example of such a push-pull driver is shown in Fig. 10-8*b*. The bias current develops two threshold drops across $Q2$ and $Q3$ that barely turn $Q5$ and $Q6$ on. When a large positive swing is needed, $Q6$ turns off and $Q5$ provides all the drive. $Q5$ and $Q6$ change roles when large negative swing is needed. Because the sources for $Q5$ and $Q6$ are not tied to their respective substrates, there is a large increase in threshold due to back-gate bias. This limits the maximum output voltage swing.

Voltage Reference

There are many applications in analog circuitry where a highly stable voltage reference is required. The absolute magnitude of this reference is often not significant because it can be scaled up or down by circuitry external to the reference. Neither is the absolute precision of the reference critical because the scaling process can include a precision trim that blows polysilicon or metal links to disconnect various trim elements. What is important though is for the reference to provide long-term stability and insensitivity to temperature and power supply variations.

A voltage reference circuit can be implemented in the standard enhancement and depletion NMOS technology by generating a voltage proportional to the difference between the enhancement and depletion threshold voltages.[6] Since the only difference between those two devices is the depletion implant, the threshold difference will be equal to Q_i/C_o, where Q_i is the implanted charge per unit area and C_o is the oxide capacitance per unit area. A circuit which senses that threshold difference is given in Fig. 10-9. Transistors $Q1$ and $Q2$ are the enhancement and depletion reference devices. By using an operational amplifier in a negative feedback mode, the output voltage V_{ref} is constrained to be at $V_{TE} + |V_{TD}|$ above ground. Furthermore, the reference devices are constrained to operate at the same drain to source voltage and the same drain current level.

By choosing a bias current level where temperature effects on gate-to-source voltage V_{GS} are minimized, and with a device size mismatch to account for a slight enhancement and depletion mobility difference, an integrated voltage reference has been built[6] that has less than 5 ppm/°C over the temperature range -55 to $125°C$.

10-3 MOS OPERATIONAL AMPLIFIERS

MOS operational amplifier design draws heavily from analogy with bipolar op-amp design. In translating design techniques from one technology to the other, one must take into consideration the lower transconductance and higher noise of MOS devices. And the gate-to-source voltage is no longer pegged at 0.7 V but could be several volts under heavy load.

CMOS Operational Amplifiers

Because of the availability of complementary devices, translation of op-amp design from bipolar is more straightforward in CMOS devices. A basic circuit[7] is shown in Fig. 10-10. The heart of the op-amp is the differential pair $P4$ and $P5$. Biasing is set up by current source $P3$ which is mirrored off the current flowing through $P1$ and R. The resistor R sets the biasing current level and can be replaced with a saturated device of the proper size. $N6$ and $N7$ are the load devices and also provide conversion of the differential signal to single ended. The differential stage consisting of $P3$, $P4$, $P5$, $N6$, and $N7$ can be reversed such that all devices switch channel type, and the load devices are connected to the positive supply, and the circuit would still function properly. But the configuration used in Fig. 10-10 provides proper level translation to feed $N9$ of the next stage. Empirical observations also show p-channel devices to have lower noise than n-channel devices of

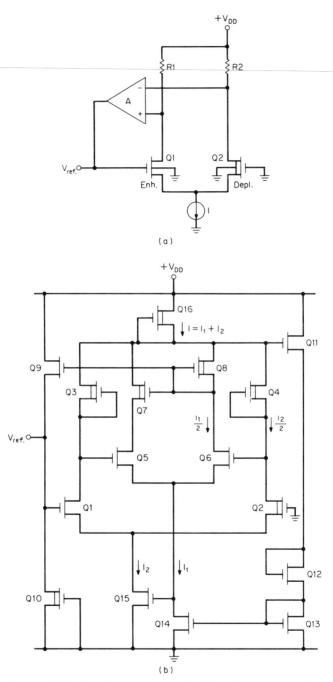

Figure 10-9 A reference voltage circuit based on sensing the threshold difference between enhancement and depletion devices. (*a*) Simplified schematic; (*b*) full schematic. (*After Ref. 6.*)

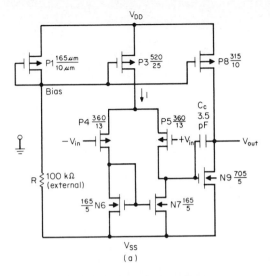

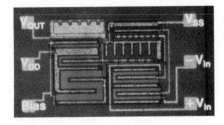

Figure 10-10 CMOS operational amplifier (without output buffer). (*a*) Schematic, (*b*) photomicrograph. (*After Ref. 7.*)

equivalent size and bias condition. And since the first stage of any cascaded amplifier should have the lowest noise so that it is not amplified by subsequent stages, it makes sense to make $P4$ and $P5$ p-channel.

The second stage comprising $P8$ and $N9$ forms an inverter to provide further amplification. The gate and drain of $N9$ represent two high-impedance modes to which the compensation capacitor C_C can be connected. The function of the capacitor can best be explained with the aid of Fig. 10-11, showing the ac equivalent circuit of a two-stage op-amp. Without C_C, the gain of the op-amp rolls off after the first pole (due to R_1 and C_1) and rolls off even further after the second pole (due to R_2 and C_2). This is illustrated in Fig. 10-12a. If the gain is still greater than 1 when the amplifier goes through a 180° phase shift, the amplifier is unstable. Addition of C_C moves the first role to a much lower frequency and the second to a much higher frequency as shown by the arrows. For this reason, C_C is referred to as a

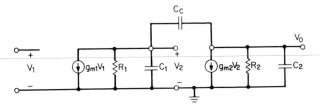

Figure 10-11 An ac equivalent circuit of a two-stage operational amplifier for analyzing pole-splitting compensation technique.

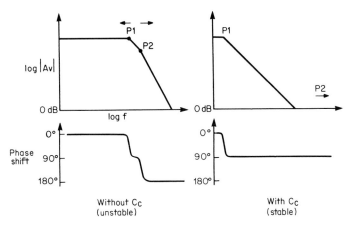

Figure 10-12 Effect of pole-splitting compensation capacitor on gain and phase shift of op-amp.

pole-splitting capacitor. The result is to force the open loop gain to drop off early enough in frequency before the phase shift reaches 180°. Because of the Miller multiplication effect, C_C need only be $\approx 10^{12}$ pF. The op-amp of Fig. 10-10 can provide gains in the range of 1000 to 100,000.

When a *small* ac signal of high enough frequency is applied to an op-amp, the output would have a reduced voltage because it is not fast enough to track the input. This is the gain fall-off at high frequency due to poles in the gain equation. However, when the output is required to go through a large voltage change, even at low frequency, it can only do so at a maximum linear rate of change called the *slew rate*. For the circuit of Fig. 10-10, the slew rate is limited by all of the input stage bias current going in to charge the compensation capacitor C_C. The slew rate is

$$(10\text{-}14) \quad \text{Slew rate} = \left.\frac{dV_{out}}{dt}\right|_{max} = \frac{I}{C_C}$$

For maximum slew rate, the compensation capacitor C_C should be reduced, consistent with keeping sufficient phase margin for stability.

NMOS Operational Amplifiers

An NMOS operational amplifier using all enhancement devices can be designed.[8] But as the analysis of Sec. 10-2, "Voltage Gain Blocks," shows, the gain that is possible is not as high as a design with depletion devices as loads.[5] A simplified circuit schematic of a NMOS op-amp with depletion devices is shown in Fig. 10-13. The input stage to the left has the two-differential input pair $M1$ and $M2$ biased with single current-source and double current-source type depletion loads. This stage provides some voltage gain. The stage to its right is a pair of low-impedance level shifters (represented as batteries) to drive a current mirror consisting of $M17$ and $M18$. This circuit provides differential to single-ended conversion while shifting the dc level of the signal to one threshold drop above V_{SS}. The second stage of gain is provided by the inverter *of* $M21$ and its current-source type load. The capacitor C_C provides the frequency compensation making this stage act as an integrator. The output buffer is designed to be a unity gain follower circuit. The implementation of the circuit blocks results in the complete circuit shown in Fig. 10-14. For a detailed explanation, the reader is referred to Ref. 5. It is clear though that the need for level translation and the lower gain per stage forces a higher transistor count in NMOS to achieve the same performance as a CMOS op-amp.

10-4 CAPACITOR-BASED CIRCUITS

Passive components available in MOS processes are resistors and capacitors. Their absolute values often vary by 10 to 50 percent, too wide a spread for

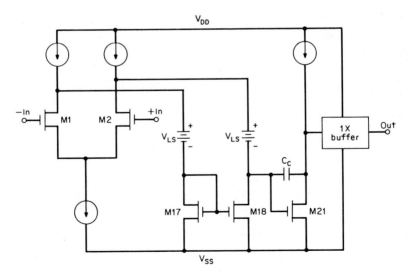

Figure 10-13 Simplified circuit schematics of NMOS op-amp using enhancement and depletion devices. *(After Ref. 5.)*

critical circuit design. As with bipolar ICs, designers turn to circuits that depend on ratios of component values. The matching characteristics of components then become of primary importance. Table 10-1 shows[9] that ion-implanted resistors match better than diffused resistors; however, capacitors give the best ratio accuracy. In addition, temperature and voltage coefficients for capacitors are substantially better. Circuits that use capacitor arrays as precision components should then have the highest accuracy and stability. For example, analog-to-digital (A/D) and digital-to-analog (D/A) circuits utilize this property for precision voltage or charge dividers to achieve 10- to 12-bit accuracy.

Certain techniques are available to minimize the ratio errors in a capacitor array. To minimize the variation due to undercutting during etching (patterning) of the capacitor plates, the perimeters must be ratioed along with the area. Figure 10-15a shows that this can be done by paralleling identical unit capacitors to form larger capacitors. A second source of error is from the long-range gradient in the oxide thickness arising from nonuniform oxide growth. Figure 10-15b shows the common centroid geometry solution that places the capacitors in such a way that they are symmetrically spaced from a common center point. A third source of error is the random edge unevenness during patterning; but the use of plasma etching can reduce

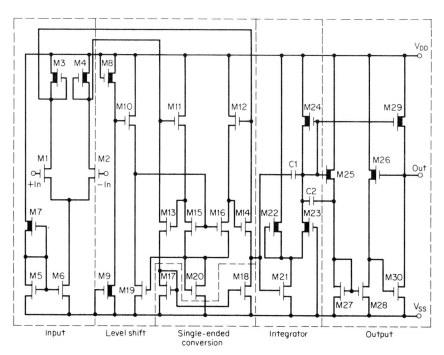

Figure 10-14 Complete circuit schematic for the op-amp in Fig. 10-13. (*After Ref. 5.*)

TABLE 10-1 Component Matching Data

Component	Fabrication technique	Matching, %	Temperature coefficient, ppm/°C	Voltage coefficient, ppm/V
Resistors	Diffused (W = 50 μm)	±0.4	+2000	~200
	Ion-implanted (W = 40 μm)	±0.12	+400	~800
Capacitors	MOS (t_{ox} = 0.1 μm $L \simeq$ 10 mils)	±0.06	26	10

Source: After Ref. 9.

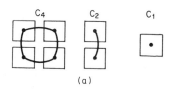

(a)

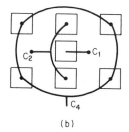

(b)

Figure 10-15 Minimizing capacitor ratio error by (*a*) scaling both area and perimeter, and (*b*) using common centroid geometry.

this effect substantially. The use of all these techniques has reduced the capacitor matching problem to such an extent that the smallest unit capacitor used is not determined by such errors but by kT/C thermal capacitor noise, parasitics, and other considerations.

10-5 SWITCHED-CAPACITOR FILTERS

To implement active filters requires very precise definition of the resistance-capacitance (RC) product. A means to provide a "resistor" with better accuracy and control than an implanted or diffused resistor is shown in Fig. 10-16a. The capacitor C is first charged to the voltage V_1. Then when the

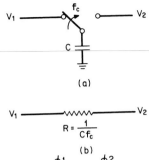

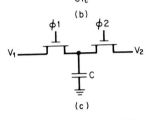

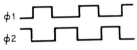

Figure 10-16 (*a*) A switched-capacitor circuit; (*b*) its equivalent resistance; and (*c*) MOS implementation and nonoverlapping clock.

switch is thrown to the right, the capacitor is discharged to voltage V_2. The amount of charge which flows from V_1 into V_2 is thus $Q = C(V_2 - V_1)$. If the switch is thrown back and forth at a clock frequency f_c, then the average current flow is

$$(10\text{-}15) \quad i = f_c C(V_2 - V_1)$$

If the clock frequency is much larger than the frequency of the signals V_1 and V_2 being sampled, then the switched capacitor of Fig. 10-16*a* is equivalent to a resistor (Fig. 10-16*b*) with a value of

$$(10\text{-}16) \quad R = \frac{1}{f_c C}$$

As an example, a 10-MΩ resistor needed for audio frequency filters can be realized by switching a 1-pF capacitor at 100 kHz rate. The MOS implementation of the switched capacitor replaces the single-pole double-throw switch with two transistors that are driven by a two-phase nonoverlapping clock ($\phi 1$ and $\phi 2$) (Fig. 10-16*c*). The overall area can be quite small since the switches are of minimum size.

A switched capacitor is used in Fig. 10-17 to replace the resistor of an RC integrator (low-pass filter). The 3-dB point of the switched-capacitor integrator is

$$(10\text{-}17) \quad \omega_o = f_c \left(\frac{C_1}{C_2} \right)$$

The response of the circuit is thus determined by capacitor *ratio* and clock frequency. As explained in the previous section, use of component ratios rather than their absolute values is the key to precise and stable circuit performance. The clock frequency is often derived from crystal-controlled oscillators, and accuracy of better than 0.1 percent is routinely achieved.

Using a switched-capacitor circuit changes the signal from continuous to sampled data. That is, an input signal is processed and appears at the output one clock cycle later. There is minimal discrepancy at the output if the clock frequency is much higher than the highest signal frequency encountered. For switched-capacitor filters, the clock frequency is often 20 to 100 times the signal frequency. Charge-coupled devices, another sampled data device, work with 8 to 10 times the clock frequency. Being sampled data devices, the design of switched-capacitor filters can draw heavily upon digital filtering techniques which also deal with elements having discrete delays.

One important aspect of using a sampled data system is the need for a prefilter to prevent *aliasing*.[10] This is the phenomenon by which high-frequency components in the signal path, particularly those higher than half the clock frequency, are reconstituted as signals in the valid frequency band (less than $f_c/2$). Figure 10-18 shows that the signal spectrum is replicated and reflected around the clock frequency and all its multiples. Conversely, if the clock frequency is too low or if there are spurious signals greater than $f_c/2$, it will alias down into the working frequency band. An antialias filter, usually a continuous low-pass filter, must then be placed in front of any sampled data device.

In switched-capacitor filter designs, there is often a need for an integrator that integrates the difference between two analog voltages. In other words, it sums (or takes the difference) between the two signals while providing the

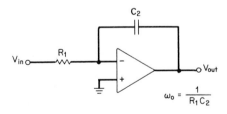

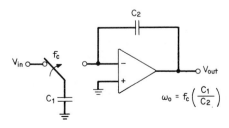

Figure 10-17 A switched-capacitor integrator whose RC time constant is a function only of capacitor ratio and clock frequency.

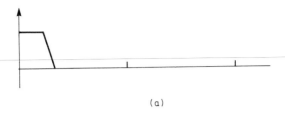

(a)

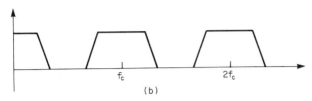

f_c $2f_c$

(b)

Figure 10-18 Signal spectrum is reflected around clock frequency and its multiple in a sampled data system. Too low a clock frequency or spurious components above $f_c/2$ will result in shifting of information down to a lower frequency band. (*a*) Desired filter response; (*b*) resulting filter response due to aliasing.

integrator function. An implementation is shown in Fig. 10-19. Its transfer function in the *s* plane is given by

(10-18) $$V_{out} = -\frac{f_c C_1/C_2}{s}(V_1 - V_2)$$

In the circuit of Fig. 10-19, as well as for any switched-capacitor circuits, the following requirements apply[11]:

1. Switched-capacitor resistors cannot by themselves close an op-amp feedback path. There is no direct path to provide continuous feedback to stabilize an op-amp.

2. There should not be any floating node for charge buildup. For example, the left plate of C_2 in Fig. 10-19 is indirectly connected to the input voltage V_1 through the switched-capacitor resistor C_1.

3. At least one plate of every capacitor must be connected to a voltage source or switched between voltage sources. Of the two plates of a capacitor, the bottom one has nonzero and nonlinear parasitic capacitance to substrate. Switching that node between voltage sources charges and discharges that parasitic capacitor but has no effect on the filter response.

4. The noninverting op-amp input should be kept at constant voltage (preferably ground). This avoids charging and discharging parasitic capacitance connected to the inverting node.

The design of switched-capacitor filters is an area under active and intense investigation. The scope of this book limits us to only one example[12] of filter design to outline the design technique and to give a flavor of the design problems and constraints.

A two-pole singly terminated low-pass *LC* ladder filter is shown in Fig. 10-20*a*. A passive *LC* ladder network shows the lowest sensitivity in fre-

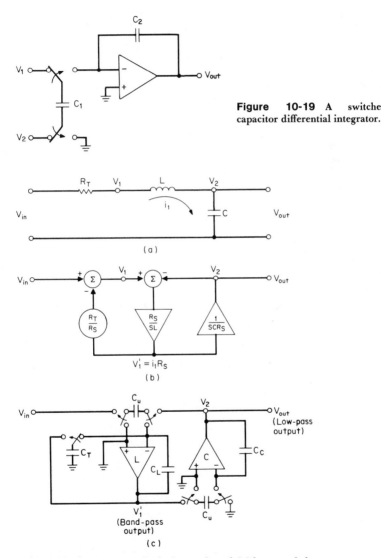

Figure 10-19 A switched-capacitor differential integrator.

Figure 10-20 (*a*) Two-pole singly terminated *LC* low-pass ladder filter; (*b*) its analog simulation; and (*c*) switched-capacitor implementation. (*After Ref. 12.*)

quency response to passive element values, particularly for a properly terminated case. For switched-capacitor filters, this translates to low sensitivity to accuracy of capacitor ratios. The values of L and C can be found from standard design tables for a normalized low-pass filter with $R_T = 1\ \Omega$ and $\omega = 1$ rad/s. Proper choice of L and C values will yield either a Butterworth filter with no ripple in the passband, or a Chebyshev filter with an equal amount of ripple in the passband, or a Bessel filter with a droopy passband and very gradual fall off, but best time delay and overshoot response. The Laplace transform equations for the circuit can be written so that they contain only integration terms:

(10-19a) $\quad V_1 = V_{in} - R_T i_1$

(10-19b) $\quad i_1 = \dfrac{1}{sL}(V_1 - V_2)$

(10-19c) $\quad V_2 = V_{out} = \dfrac{1}{sC} i_1$

Since the filter will be implemented with op-amps which are voltage-controlled sources, one can write a set of all-voltage equations by multiplying i_1 by a scale resistance such that $V_1' = i_1 R_S$. Thus

(10-20a) $\quad V_1 = V_{in} - \dfrac{R_T}{R_S} V_1'$

(10-20b) $\quad V_1' = \dfrac{R_S}{sL}(V_1 - V_2)$

(10-20c) $\quad V_2 = V_{out} = \dfrac{1}{sR_S C} V_1'$

Figure 10-20b is a schematic representation of the circuit showing the integration function with triangles, scaling function with circle, and addition (or subtraction) of voltages with summers.

The switched-capacitor implementation of the filter is obtained by replacing the conventional integrators with their switched-capacitor equivalents using the differential integrator of Fig. 10-19. The result is shown in Fig. 10-20c. Note that the differential integrator provides the summing function also. Comparing the constant of proportionality among Eqs. (10-18) and (10-20b) and (10-20c) yields

(10-21a) $\quad \dfrac{C_L}{C_u} = \dfrac{Lf_c}{R_S}$

(10-21b) $\quad \dfrac{C_C}{C_u} = CR_S f_c$

The termination path R_T/R_S, implemented by the capacitor C_T, simply scales through the integrator in the center of Fig. 10-20b. By analogy with Eq. (10-21a), one obtains

$$(10\text{-}21c) \quad \frac{C_L}{C_T} = \frac{Lf_c}{R_T}$$

The scaling resistor R_S can be used to adjust the capacitor ratios C_L/C_u and C_C/C_u. Since it increases one while decreasing the other, the minimum total area is when $C_L/C_u = C_C/C_u$, which implies that

$$(10\text{-}22) \quad R_S = \left(\frac{L}{C}\right)^{1/2}$$

All capacitor ratios can then be determined.

Other aspects of switched-capacitor filter design are to scale individual filter sections such that all op-amp outputs have the same peak voltage (to avoid saturating one before the other), to minimize overall noise, and to compensate for slight warping of frequency response due to finite clocking frequency. There are also numerous design techniques different from the one described above, using primarily the z transform to represent digital filtering techniques. The interested reader is referred to the literature for further details.[12,13]

10-6 CHARGE-COUPLED DEVICES

A charge-coupled device (CCD) is an array of closely spaced ($< 1 \ \mu$m) MOS capacitors.[14] By sequentially pulsing sets of electrodes, packets of inversion-layer charge representing the signal can be propagated down the array. The voltages applied to the gate electrodes are greater than the threshold voltage, while the inversion-layer charge is less than the equilibrium value so that each MOS capacitor is always in deep depletion (nonequilibrium). Figure 10-21 illustrates this principle. With positive gate voltage V_G applied, the devices goes into deep depletion, resulting in a large surface potential of ϕ_S with its attendant large band bending. When the signal charge is shifted in from an adjacent capacitor, the surface potential is reduced, resulting instead in a larger field across the oxide. The electrodes have to be closely spaced in order for the depletion layers under them to couple together and allow the signal charge to transfer completely.

The quantity ϕ_S, the surface potential (voltage drop across the silicon), is used repeatedly in CCD work. To solve for it, Eq. (3-12) from Chap. 3 is again invoked:

$$(3\text{-}12) \quad V_G = \frac{-Q_B}{C_o} + \phi_S = \frac{\sqrt{2\epsilon_s\epsilon_o q N_A \phi_S}}{C_o} + \phi_S$$

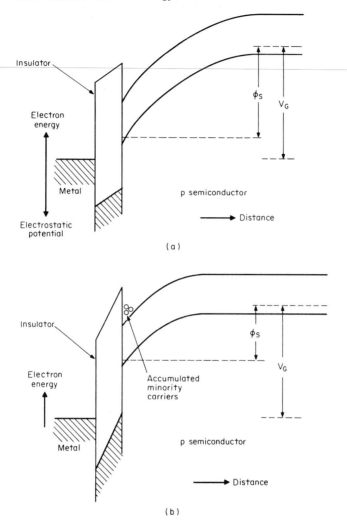

Figure 10-21 Energy-band diagram of a MOS capacitor (*a*) with gate voltage applied and no signal charge, (*b*) with signal charge reducing the surface potential. (*After Ref. 15.*)

By defining a constant[15]

$$(10\text{-}23) \quad V_0 = \frac{qN_A\epsilon_s\epsilon_o}{C_o^2}$$

one can arrive at the expression (see Prob. 6)

$$(10\text{-}24a) \quad \phi_S = V_G + V_0 - (2V_GV_0 + V_0^2)^{1/2}$$

or

(10-24b) $$\phi_S = V_G - V_0\left[\left(1 + 2\frac{V_G}{V_0}\right)^{1/2} - 1\right]$$

With the nonzero flatband voltage, or with the presence of signal charge, V_G in Eq. (10-24) is replaced by the effective gate voltage $V_G' = V_G - V_{FB} + Q_n/C_o$, where Q_n is in units of coulombs per square centimeter and should algebraically be negative if electrons are the signal charge. Sample plots of Eq. (10-24) are given in Figure 10-22.

The signal charge density Q_n, though in units of coulombs per unit area, has an actual charge distribution with a well-defined peak value of n_s carriers per unit volume at the surface and drops off sharply in about 50 to 100 Å. The relationship between n_s and Q_n can be shown[16,17] to be

(10-25) $$n_s = \frac{WQ_n}{qL_D^2} + \frac{1}{2N_A}\left(\frac{Q_n}{qL_D}\right)^2$$

where W is the depletion width, and L_D is the *extrinsic* Debye length:

$$L_D = \sqrt{\frac{\epsilon_s\epsilon_o kT}{q^2 N_A}}$$

Figure 10-23 is a graph of Eq. (10-25) for some typical parameters.

The operation of a CCD is best illustrated by the three-phase system of Fig. 10-24. A p-type substrate is used, producing an n-channel device whose signal charge is composed of electrons. The clocking signals $P1$, $P2$, and $P3$ are positive pulses and biased to be constantly greater than the threshold voltage. The signal charge initially collects under the electrode $P1$, that having the most positive potential ($t = t_1$). It then transfers over to the adjacent

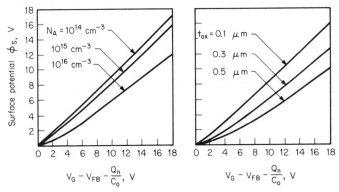

Figure 10-22 Surface potential versus effective gate voltage for (a) various substrate dopings with 1000-Å gate oxide, (b) various oxide thicknesses with acceptor doping of 10^{15} cm^{-3}.

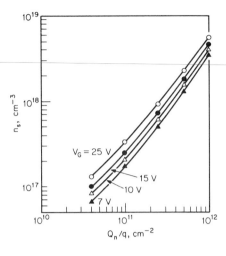

Figure 10-23 Relationship between electron volume concentration at the surface and surface charge density in a CCD for various applied gate voltages. $t_{ox} = 1200$ Å; $N_A = 10^{15}$ cm^{-3}. (*After Ref. 16.*)

electrode when the most positive potential moves to $P2$ ($t = t_4$). A three-phase system is required to isolate charge packets and still provide directionality in shifting. With a three-phase system, the charge packet has to transfer past three electrodes to move through one shift register bit.

The signal charge reduces the surface potential almost in direct proportion (see Fig. 10-22 and Prob. 7), leading to the popular use of the fluid analogy where charges fill up the potential "wells." However, the signal charge physically always travels at the oxide-semiconductor interface and hence is termed a *surface-channel* device. Note in Fig. 10-24 that a heavy p^+ diffusion (called channelstop) at the edge of the CCD confines the charge in that lateral direction. The heavy p^+ raises the substrate doping such that the surface potential is near zero, forming a potential barrier.

By the use of a thin epitaxial or ion-implanted layer whose polarity is opposite to that of the substrate (Fig. 10-25a), the potential maximum (energy-band diagram minimum) will move away from the silicon surface. The charge will physically travel in the silicon bulk. It will be shown shortly that such a device, called a *buried-channel CCD,* has better performance than a surface-channel CCD. It is indicated in Fig. 10-25a that a large reverse bias needs to be applied to the layer through the input and output diodes to drain away all mobile carriers until the layer is fully depleted.

To use two-phase clocks and still provide directionality, a gradient in the surface potential profile must be provided. Figure 10-26 shows how such a gradient is obtained with a set of four electrodes per bit, connected in pairs. For each of the pairs, the same gate voltage produces a difference in surface potential due to the difference in oxide thicknesses. A similar potential difference could be obtained if the oxide thicknesses were the same but the substrate doping concentrations were different, such as by ion implantation.

There are two major limitations to the performance of CCDs, namely transfer inefficiency and the thermal carrier generation (dark current). The former limits the high-frequency operation of practical devices to about 10 MHz, while the latter limits the low-frequency operation to around 1 kHz.

Transfer inefficiency results from the fact that not all the carriers in a charge packet can transfer from one electrode to the next within a clock period. The fraction that is left behind *per transfer* is the transfer inefficiency ϵ. A typical value of ϵ is 10^{-4} per transfer.

The transfer of carriers is not a linear time function but a result of three basic charge-transfer mechanisms. At the beginning, the transfer is dominated by self-induced drift produced by electrostatic repulsion of the carriers amongst themselves. When a small amount of signal charge is left, the transfer mechanism is primarily thermal diffusion. In this case, the charge under

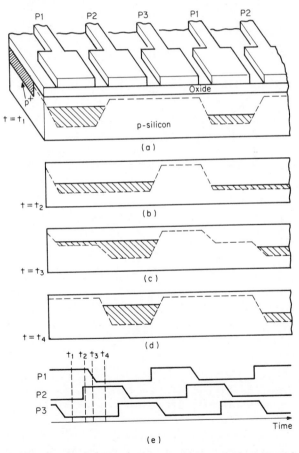

Figure 10-24 Operation of a three-phase surface *n*-channel CCD. The charge-carrying potential wells are drawn on the substrate cross section. (*After Ref. 15.*)

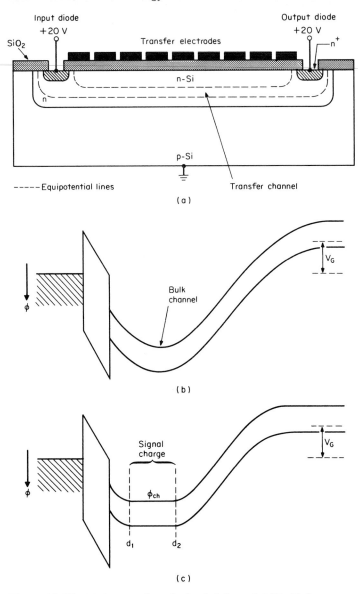

Figure 10-25 (*a*) Cross section of a buried-channel CCD; (*b*) its energy-band diagram without signal charge, and (*c*) with signal charge. The potential minimum has moved away from the oxide-semiconductor interface.

the storage (sending) electrode decreases exponentially with time with a time constant that is proportional to the square of the electrode length. Finally, the last few remaining carriers respond primarily to the fringing field that is set up by the gradient in surface potential between adjacent electrodes. Buried-channel devices result in higher fringing fields and hence can achieve much higher transfer efficiency. Transfer efficiency can be improved by using shorter gate lengths, lightly doped substrate, and closely spaced electrodes.

Thermal carrier generation results from the fact that the device is constantly in deep depletion. In an n-channel device, holes from electron-hole pairs generated in the depletion region are driven toward the substrate, while electrons collect at the surface, adding to the signal charge. In digital applications, more charge is added to a 0 (no charge) than to 1 (with charge), reducing the detectable difference between the two.[18] In analog applications, the added charge reduces the dynamic range, whereas in imaging applications, the added charge produces background noise (hence the term *dark current*) which is particularly undesirable because it is not uniform across the device. Dark current can be reduced by process changes that increase minority carrier lifetime and reduce surface-state density.

10-7   CHARGE-COUPLED DEVICE APPLICATIONS

A CCD is essentially a serial device. Its basic structure is a shift register consisting of a linear array of electrodes with input and output diodes and control electrodes at the ends.

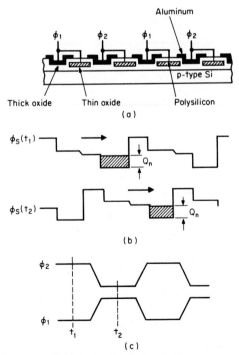

Figure 10-26   A two-phase system producing a gradient in surface potential by means of oxide thickness differences.

Introducing signal charge is by means of the "fill and spill" technique. One such implementation for sampling an analog signal is shown in Fig. 10-27. The clocking waveforms of Fig. 10-27b are referenced to ground, but a negative bias is applied to the substrate to bias all voltages greater than threshold. Allowing clock voltages to drop below threshold will collapse the depletion region, causing the signal charge to recombine with majority carriers. Electrode $G1$ is tied to $\phi1$, while the analog input is applied to $G2$. The voltage at $G2$ should be sampled by $\phi1$, that is, shorted to ground by transmission gates whenever $\phi1$ is not 1. As shown in Fig. 10-27c, when V_S goes low at t_2, charges fill under both electrodes $G1$ and $G2$. But when V_S returns high at t_3, all excess charge spills back out and only the charge represented by the surface potential difference under $G1$ and $G2$ is captured and transferred forward in subsequent clock cycles. For digital applications,

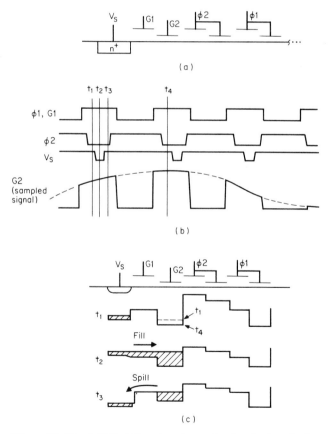

Figure 10-27 A CCD shift register input circuitry for introducing signal charge by sampling an analog input. (*a*) The device cross section; (*b*) clocking waveforms; and (*c*) surface potential profiles.

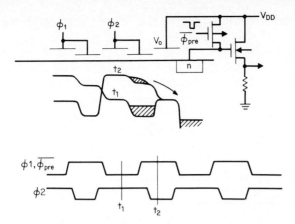

Figure 10-28 A CCD shift register output circuitry. (*a*)
Device cross section with surface potential profile; and (*b*)
clocking waveforms.

$G2$ is also tied to $\phi1$ and another electrode is added in front to $G1$ which
can permit or inhibit the flow of carriers from the input diode.

The output structure, common to both analog and digital devices, is
shown in Fig. 10-28. A constant potential barrier is formed under electrode
V_O to contain the signal charge at t_1. When $\phi2$ goes low at t_2, the charge is
pushed over the barrier and drained by the output diode. The output node
is precharged to a fixed positive dc level. The signal charge will then cause
a negative change in the output node voltage by an amount equal to the
product of the signal charge and the output node capacitance. As shown, a
source follower is needed to buffer the output node. A properly designed
device can have output voltage swings of up to several volts.

The first use of a CCD was as a delay line whose delay time was variable
with the clock frequency. Early development efforts were also directed
toward a high-density serial memory. The memory organization is usually
a serial-parallel-serial (SPS) format. Data are read in at full clock rate f_c
into an input register m bits long. After a line of data is loaded in, it is
transferred in parallel to an array of m registers, each n bits long. All data
move through the main $m \times n$ array at the much slower clock frequency
of f_c/m. At the output, a similar serial register m bits long accepts data in
parallel and shifts them out in parallel at full f_c rate. Since the main array
uses a much lower frequency, less power is consumed. Just as important,
each charge packet moves through only $p(m + n)$ transfers, where p is the
number of clock phases, improving the overall transfer efficiency. High-density memory development has shifted from CCDs to dynamic RAMs
because cost reduction due to application of scaling laws has not been as
successful with CCDs.

Another application for the CCD is as a solid-state self-scanning image sensor to replace electron-beam-scanned camera tubes.[19] Both linear and area arrays are possible. Photon energy, after penetrating the semitransparent polysilicon electrodes, generates electron-hole pairs in the depletion layers. After an integration time of several hundred milliseconds, the collected electrons are shifted out at very high speed. Techniques are available to prevent charges generated from points of intense illumination from "blooming" out to adjacent areas. Color sensors are created by splitting the incoming beam optically and using three different filter and CCD sets.

The other major area of CCD application is in signal processing. The inherent sampled-analog, discrete time-delay nature of CCDs is well utilized in the construction of transversal filters. Transversal filters form a subset of a general class of digital filters whose output is never fed back to the input and thus are finite impulse response (FIR) filters. The other class of filters is the recursive filters where the output signal is added to the input, forming an infinite impulse response (IIR) filter. With this feedback path, a given filter function can be synthesized with a much smaller number of delay stages, such as in a switched-capacitor filter. However, this form is generally avoided with CCDs because of severe sensitivity to the amount of feedback.

How a transversal filter works can best be understood with the aid of Fig. 10-29a, where an input signal is shown being sampled and passed through a series of discrete time delays. The signal is sampled after every delay, multiplied by the appropriate tap weights h_k and summed. The output after n clock periods is

(10-26) $\quad V_{\text{out}}(nT_c) = \sum_k h_k V_{\text{in}}(nT_c - kT_c)$

With a large number of taps, Eq. (10-26) approximates

(10-27) $\quad V_{\text{out}}(t) \cong \int_0^t h(\tau) V_{\text{in}}(t - \tau)\, d\tau$

Equation (10-27) describes the convolution of the two functions in the time domain.

(10-28) $\quad V_{\text{out}}(t) = h(t) \otimes V_{\text{in}}(t)$

That is,

(10-29) $\quad V_{\text{out}}(f) = H(f) V_{\text{in}}(f)$

It is interesting to note that the tap weights h_k actually describe the impulse response of the filter $H(f)$.

The implementation of the weighting and summing function is by means of the split-electrode technique.[20] As shown in Fig. 10-29b, the electrodes of one of the clock phases are split in the ratio $(1 + h_k):(1 - h_k)$ so as to

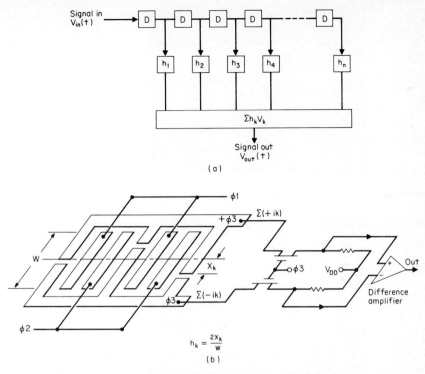

Figure 10-29 A CCD transversal filter. (*a*) Digital representation of filter; (*b*) implementation with split-electrode technique.

produce a net tap weight of h_k at that node. For example, a tap weight of 0.0 will result in the split at the middle, and a normalized tap weight of $+1.0$ will result in the split at one end. In clocking an electrode, the charging current is proportional to the signal charge under the electrode as well as its area. Thus the difference in charging current to the two halves of the split is the sum of the products between the signal at a node and the tap weight for that node. This difference is then the desired output signal.

In practice, the output signal is obtained by passing the charging currents through resistors and sensing the difference through an on-chip operational amplifier.

The results for a 64-tap bandpass and low-pass tranversal filter are shown in Fig. 10-30. The filters were fabricated on a CCD/CMOS process[7] with the n-channel CCD inside the p-tub. The filters show in-band ripples of less than 0.6 dB and stop-band ripples of 28 to 32 dB.

As with switched-capacitor filters, the frequency responses of CCD filters are dependent only on geometrical factors and hence are very stable and reproducible. The response, in fact, scales with clock frequency. Unlike switched-capacitor filters, CCDs have a relatively low clock frequency, usually 8 to 10 times the filter corner frequency.

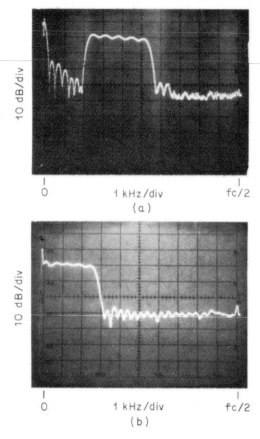

Figure 10-30 Frequency response of a 64-tap CCD transversal filter (*a*) bandpass, and (*b*) low-pass. (*After Ref. 7.*)

A detailed comparison of switched-capacitor and CCD filters can be found in Refs. 21 and 22.

REFERENCES

1. P. R. Gray, D. A. Hodges, and R. W. Brodersen (eds.), *Analog MOS Integrated Circuits,* IEEE Press, New York, 1980, pp. 28–49. *Note:* This book contains reprints of many papers concerning analog MOS circuits, including many of the references in this chapter.

2. R. S. Ronen, "Low-frequency 1/f noise in MOSFETs," *RCA Review,* vol. 34, pp. 280–307, June 1973.

3. H. Mikoshiba, "1/f noise in n-channel silicon-gate MOS transistors," *IEEE Trans. Electr. Dev.,* vol. ED-29, no. 6, pp. 965–970, June 1982.

4. Y. P. Tsividis, "Design considerations in single channel MOS analog integrated circuits—a tutorial," *IEEE J. Solid-State Circ.,* vol. SC-13, pp. 383–391, June 1978.

5. D. Senderowicz, D. A. Hodges, and P. R. Gray, "High performance NMOS operational amplifiers," *IEEE J. Solid-State Circ.,* vol. SC-13, pp. 760–766, December 1978.

6. R. A. Blauschild et al., "A new NMOS temperature-stable voltage reference," *IEEE J. Solid-State Circ.,* vol. SC-13, no. 6, pp. 767–774, December 1978.

7. D. G. Ong, "An all-implanted CCD/CMOS process," *IEEE Trans. Electr. Dev.,* vol. ED-28, no. 1, pp. 6–12, January 1981.

8. Y. P. Tsividis and P. R. Gray, "On integrated NMOS operational amplifier with internal compensation," *IEEE J. Solid-State Circ.,* vol. SC-11, pp. 748–754, December 1976.

9. D. A. Hodges, P. R. Gray, and R. W. Brodersen, "Potential of MOS technologies for analog integrated circuits," *IEEE J. Solid-State Circ.,* vol. SC-13, no. 3, pp. 285–294, June 1978.

10. L. R. Rabiner and B. Gold, *Theory and Application of Digital Signal Processing,* Chap. 2, Prentice-Hall, Englewood Cliffs, N.J., 1975.

11. R. W. Brodersen, P. R. Gray, and D. A. Hodges, "MOS switched-capacitor filters," *Proc. IEEE,* vol. 67, no. 1, pp. 61–75, January 1979.

12. G. M. Jacobs et al., "Design technique for MOS switched capacitor ladder filters," *IEEE Trans. Circ. Sys.,* vol. CAS 25, no. 12, pp. 1014–1021, December 1978.

13. G. C. Temes, H. J. Orchard, and M. Jahanbegloo, "Switched-capacitor filter design using the bilinear z-transform," *IEEE Trans. Circ. Sys.,* vol. CAS 25, no. 12, pp. 1039–1044, December 1978.

14. G. Hobson, *Charge-Transfer Devices,* Chap. 1, Wiley, New York, 1978.

15. C. H. Séquin and M. F. Tompsett, *Charge Transfer Devices,* Chap. 2, Academic, New York, 1975.

16. D. G. Ong and R. F. Pierret, "Approximate formula for surface carrier concentration in charge-coupled devices," *Electr. Lett.,* vol. 10, no. 1, pp. 6–7, Jan. 10, 1974.

17. S. M. Sze, *Physics of Semiconductor Devices,* 2d ed., Chap. 7, Wiley, New York, 1981.

18. D. G. Ong and R. F. Pierret, "Thermal carrier generation in charge-coupled devices," *IEEE Trans. Electr. Dev.,* vol. ED-22, no. 8, pp. 593–602, August 1975.

19. D. F. Barbe, "Imaging devices using the charge-coupled concept," *Proc. IEEE,* vol. 63, no. 1, pp. 38–67, January 1975.

20. R. D. Baertsch et al., "The design and operation of practical charge-transfer transversal filters," *IEEE Trans. Electr. Dev.,* vol. ED-23, pp. 133–141, February 1976.

21. R. W. Brodersen and T. C. Choi, "Comparison of switched-capacitor ladder and CCD transversal filters," *Proc. 5th Int. Conf. CCD,* Edinburgh, pp. 268–278, 1979.

22. C. R. Hewes, R. W. Brodersen, and D. D. Buss, "Applications of CCD and switched-capacitor filter technology," *Proc. IEEE,* vol. 67, pp. 1403–1415, October 1979.

PROBLEMS

1. To calculate for λ, the output resistance parameter, let

(10-30) $$I_D = \frac{\beta_o}{2} \frac{W}{L - x_d} (V_G - V_T)^2$$

where the channel shortening is due simply to a depletion widening of the drain junction. That is,

(10-31) $$x_d = \sqrt{\frac{2\epsilon_s\epsilon_o(V_D + V_B)}{qN_A}}$$

where V_D is the applied drain voltage, and V_B is the junction built-in potential. Derive the output conductance whence the parameter λ is obtained.

(10-32) $$g_o = \lambda I_o = \frac{\partial I_D}{\partial V_D}$$

2. What are the characteristics of the CMOS output buffer of Fig. 10-8*b* with regard to output swing, output impedance, and voltage gain? What would be the effect of resistive load on the output? How do those same characteristics compare with the class B common-source output circuit of Fig. 10-31?

3. Draw a source follower circuit that has a resistor as load. Assume the source is tied to the substrate to neglect back-bias effects. Plot the $V_{in} - V_{out}$ transfer characteristics if $R = 10$ kΩ and $\beta = 10$ μA/V^2. What is the voltage gain at $V_{out} = 5$ V?

4. Sketch the cross section of capacitor structures available in
   (a) a metal-gate process
   (b) a single-polysilicon gate process
   (c) a double-polysilicon gate process

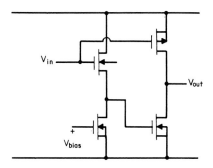

Figure 10-31 A class B common-source CMOS output buffer circuit (Prob. 2).

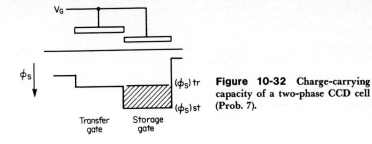

Figure 10-32 **Charge-carrying capacity of a two-phase CCD cell (Prob. 7).**

5. What would be the effect of the gate-to-source and drain overlap capacitance when the clock for a switch-capacitor circuit turns off the switching transistors? What might be a solution to minimize that effect?

6. Starting with the loop equation of Eq. (3-12), and defining the constant V_0 in Eq. (10-23), solve the quadratic equation for ϕ_S that results, to arrive at Eq. (10-24). Use the boundary condition $\phi_S = 0$ when $V_G = 0$ to pick the proper sign for the square root term.

7. The cross-hatched area in Fig. 10-32 represents the maximum charge-carrying capacity for a dual oxide two-phase CCD cell. Derive $Q_{n,\max}$ expressed in terms of the surface potential under the storage gate and under the transfer gate.

11

CMOS

11-1 ADVANTAGES AND DISADVANTAGES OF CMOS

Complementary MOS is increasingly specified as the MOS technology to use, particularly as integrated circuits proceed toward larger and larger scales of integration.[1] Its advantages over single-channel devices are as follows:

1. *Low-Static Power Dissipation* When no node voltages are changing, CMOS has extremely low power dissipation. The availability of complementary devices permits one of the devices in an inverter or logic gate always to be off. The only residual current drawn is from leakage current, typically ≈ 1 μA. The operating power does increase with frequency and causes considerable current to be drawn when ≈ 1 MHz; still, CMOS circuits end up with lower total power dissipation than NMOS even at a high operating frequency because only a small section of the chip is operating at maximum speed.

2. *Wide Power Supply Range* The most popular version of CMOS circuits can operate from 3 to 15 V. This compares with transistor-transistor logic (TTL) circuits that operate from 5 V $\pm$ 10 percent. CMOS has the unique property that the switching point of its inverters (and logic gates) scales with supply voltage. The noise margin, as a percentage of the supply voltage, is thus preserved.

3. *Higher Speed* Properly designed CMOS circuits can operate at 10 to 25 MHz or higher, matching ECL (emitter-coupled logic) speeds.[2] This is possible because load and driver devices both have large drive capability.

4. *Wide Output Voltage Swing* The output can swing from ground to the positive supply rail without suffering from threshold drops or having to resort to bootstrap circuits.

CMOS does require more complex processing (estimated to take 20 percent more steps), more layout area (5 to 20 percent higher), and has the possibility of latch-up. All of the above items will be explained in further detail in this chapter.

11-2 CMOS CIRCUITS

Inverter

The basic CMOS circuit is an inverter, as shown in Fig. 11-1. An n-channel device serves as the "driver," while a p-channel device serves as the "load." The two devices have a common gate and common drain. Their body terminals are tied to their respective sources. (The term *body* will be used in this chapter for the lead that was previously referred to as substrate, to avoid confusion with the physical substrate, i.e., the wafer.)

The foremost attribute of a CMOS inverter is that there is no dc flow when the inverter is not changing state. When the input is a 1 (V_{DD}), the n-channel device is turned on, but the p-channel device is off (its $V_{GS} = 0$), so there is no dc path from V_{DD} to V_{SS} (ground). When the input is at a 0 (ground), the p-channel device is on, but the n-channel device is off; so again there is no direct current flow. The only current that flows is leakage current of reverse-biased junctions. There is significant current flow, however, when the inverter is changing states at a rapid rate. This will be discussed specifically in Sec. 11-4, "Power Dissipation."

The cross section of a CMOS inverter is shown in Fig. 11-2. The need for a different type substrate for the n-channel device is satisfied by the use of a deep p-tub (or p-well) diffusion. With the output node constrained to swing from ground to V_{DD}, the drain-to-body junctions are always reverse-

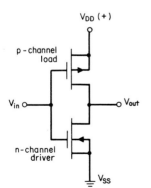

Figure 11-1 CMOS inverter.

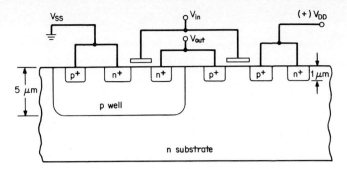

Figure 11-2 Cross section of a CMOS inverter.

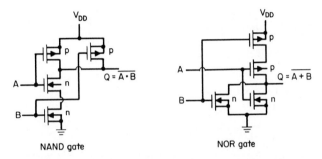

Figure 11-3 Two-input CMOS NAND gate and NOR gate.

biased. The p-tub–to–substrate junction is also always in reverse bias, having the power supply voltage across it. Appropriate diffusions are needed to form ohmic contacts to p-tub and n substrate. The power supply and ground connections are then brought up from the top of the die, and there is no compelling reason to bond to the back of the wafer. This is in contrast to NMOS circuits, where the only contact to the body is via the gold-alloyed back contact, and the header to which the die is attached must have an electrical connection brought out.

CMOS Logic Gates

A CMOS two-input NAND gate and NOR gate are shown in Fig. 11-3. To ensure that there is no dc path during quiescent states, a pair of complementary devices must be provided for each input. CMOS circuits thus have a higher transistor count than NMOS, which adds one device for every input after the first. Going beyond a simple device count, CMOS logic gates also always have two or more devices in series. To maintain equivalent current drive capability, each of the devices in series must be scaled up by a factor equal to the number of devices. This too increases layout area. Fur-

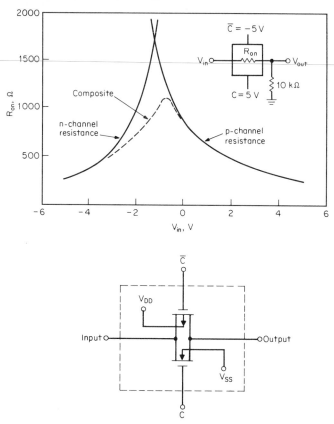

Figure 11-4 CMOS transmission gate and its series resistance. *(After Ref. 3. Used with permission of McGraw-Hill Book Company.)*

thermore, to prevent latch-up, CMOS circuits are laid out with guard rings surrounding the devices. Guard rings are diffusions of the same polarity as the body of the device.

Transmission Gates

A well-utilized circuit element that is unique to CMOS is the transmission gate as shown in Fig. 11-4. It consists of p-channel and n-channel devices connected in parallel. Opposite polarity clock signals C and $\overline{C}$ are applied to the gates such that both devices turn on and off together. When either device is used by itself as a transmission gate, its series resistance increases dramatically as its source and drain voltages approach the clock voltage[3] (Fig. 11-4). By having the two channel types in parallel, one device has a resistance that decreases whenever the other one is increasing. Shown in the figure is the resulting net resistance.

CMOS Static RAM Cell

A fully static CMOS RAM cell is shown in Fig. 11-5. The cell has six transistors per cell, the same count as the NMOS version; yet it has an extremely low static power dissipation and a large capability to drive the bit lines in both directions.

11-3 CMOS CIRCUIT ANALYSIS

CMOS circuits are often designed to be run with a wide power supply range. The transfer characteristics for various power supply voltages are shown in Fig. 11-6. It can be seen that the switching points (the intersection of the $V_{in} = V_{out}$ line with the various curves) are always at roughly the

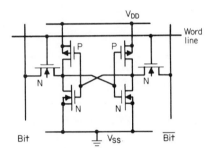

Figure 11-5 CMOS static RAM cell.

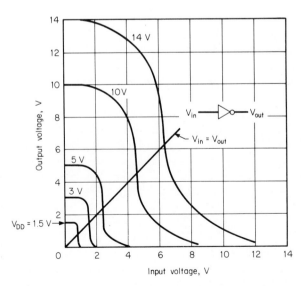

Figure 11-6 Transfer characteristics of various power supply voltages.

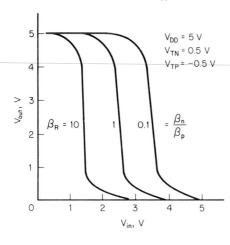

Figure 11-7 Transfer characteristics for various ratios of *n*- to *p*-channel β.

same fraction of the power supply voltage. Specifically, the transition voltage, or the switching point, is given by

$$(11\text{-}1) \quad V_{\text{transition}} = V_{\text{in}} = V_{\text{out}} = \frac{V_{DD} + V_{TP} + V_{TN}\,(\beta_N/\beta_P)^{1/2}}{1 + (\beta_N/\beta_P)^{1/2}}$$

Equation (11-1) can be derived (see Prob. 1) by noting that when $V_{\text{in}} = V_{\text{out}}$, both devices have their gates effectively tied to their drains. The two saturation currents can then be equated to each other. Notice that when $V_{TP} = -V_{TN}$ and $\beta_N/\beta_P = 1$, $V_{\text{transition}} = V_{DD}/2$. Then both the 1 and 0 noise margins are equally maintained at 50 percent of the power supply voltage.

The *n*- and *p*-channel device gain and threshold voltage tend to track with temperature with the result that the transfer curves of Fig. 11-6 change very little with temperature.

Equation (11-1) also suggests a way to skew the transition voltage purposely away from the $V_{DD}/2$ point to suit specific applications. Figure 11-7 shows the transfer characteristics for several ratios of *n*- to *p*-channel β. For example, with a β ratio of 10, it is suitable for 5-V CMOS input circuits to interface with TTL circuits whose switching level is 1.2 V. One must be cautioned that in driving CMOS circuits, if the input levels are maintained at quiescent levels near the transition voltage, significant dc will flow, defeating one purpose of using CMOS.

11-4 POWER DISSIPATION

The power dissipation of CMOS circuits has a number of components that have varying importance depending on the situation.

Static Power Dissipation

The static power dissipation of CMOS circuits is extremely small, often $\approx$ 10^{-6} W. The power drain consists only of leakage current. Figure 11-8 shows the path of leakage current I_L for an inverter whose input is at a 1 and the n-channel device is turned on. (Results are analogous when the input is a 0.) One path of I_L is through the reverse-biased drain junction (in this case, of the p-channel device) and then through the turned on n-channel device. The other path is through the reverse-biased p-tub to substrate junction. The tub leakage component is usually larger because of the larger tub area and larger depletion width.

The static power dissipation is

$$(11\text{-}2) \quad P_S = I_L V_{DD}$$

Dynamic Power Dissipation

The dynamic power dissipation is given by

$$(11\text{-}3) \quad P_D = CV_{DD}^2 f$$

where C is the load capacitance seen by the inverter, and f is the frequency of operation. A heuristic basis for Eq. (11-3) comes from the observation that the output capacitance has a stored energy of $\tfrac{1}{2}CV_{DD}^2$. The power expended to charge and discharge that energy with a frequency f is given by Eq. (11-3). At frequencies $\approx$ 1 MHz, this component can be quite significant (see Prob. 2). It becomes comparable to TTL implementations.

Device Dissipation

There is another component of power dissipation that takes place when the input voltage changes states. As illustrated by Fig. 11-9, considerable cur-

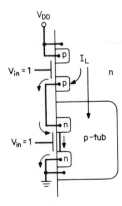

Figure 11-8 Leakage paths that determine static power dissipation.

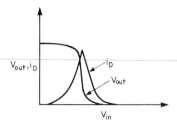

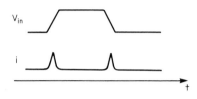

Figure 11-9 Direct current flow through an inverter.

rent flows when the input is between a full 0 and full 1 and both *p*- and *n*-channel devices are on. This component is described by

$$(11\text{-}4) \quad P_d = \int_{1 \text{ cycle}} V_{\text{out}} i \; dt$$

When the rise- and falltimes of the input waveform are small, this component becomes negligible. Furthermore, when there is capacitance loading such that the output voltage is still near either power supply rail when both devices are on, then the current flow remains small. For this reason, the power dissipation of CMOS circuits under switching conditions is caused almost entirely by capacitive loading.

11-5 PROCESSING ISSUES

One issue while developing a CMOS process is whether to use a *p*-tub in an *n*-type substrate or an *n*-well in a *p*-type substrate.

The *p*-tub process was a natural outgrowth of the PMOS technology. It uses the same *n*-type wafers, and the less dominant *n*-channel devices were contained in a separately introduced *p*-tub. When metal-gate CMOS was prevalent, the *p*-tub was always doped by diffusion, making the *p*-tub doping always higher than the substrate. It is fortuitous that *n*-channel devices need a higher doping concentration to remain in enhancement mode (see Sec. 7-2).

With ion implantation and the silicon gate process available, the previous constraints no longer apply. The background doping concentrations both inside and outside the wells can be independently adjusted. Threshold voltages can also be adjusted at will. Having an *n*-well on a *p* substrate then becomes practical.[4] Since certain circuits such as memories tend to have a higher percentage of n-channel devices, an n-well process allows direct translation of those circuits from HMOS or NMOS to CMOS. p-Channel devices are then used primarily in the periphery input-output circuits. Placing p-channel devices inside heavily compensated wells might tend to lower their mobility, accentuating the mobility difference with n-channel devices. However, whatever degradation is observed tends to be small enough to be insignificant.

In linear circuits, the implication of an n-well process is that it is the p-channel devices that can have their body tied to their source. In certain circuit designs, such a connection is used as a technique to reduce the threshold voltage drop across output buffers and improve power supply rejection.

Whether a p-tub or n-well is used, it is good layout practice to merge the wells of adjacent devices together. Liberal placement of contacts to the well will prevent latch-up by reducing voltage drops caused by parasitic currents flowing in the well.

A second issue with CMOS processing is to quantify its increased processing complexity. Table 11-1 compares the masking steps between the NMOS process and the standard CMOS silicon gate. Note that for certain steps such as source and drain implants, an unmasked operation in NMOS requires two masking steps in CMOS. The result is a significantly longer flow for CMOS. However, as process flows become more complex, CMOS requires only a few additional steps to achieve the same performance. This

TABLE 11-1 Comparison of Masking Steps between CMOS and NMOS, and High-Performance CMOS and HMOS

Basic flow		High performance	
NMOS	**CMOS**	**HMOS**	**CMOS**
	p-Tub		p-Tub
Active area	Active area	Active area	Active area
	n-Channel field	Depletion implant	
Poly gate	Poly gate		n-Channel field
	p^+ Source and drain	Buried contact	Buried contact
	n^+ Source and drain	Poly gate	Poly gate
Contacts	Contacts		p^+ Source and drain
Metal	Metal		n^+ Source and drain
Passivation	Passivation	Contact	Contact
		Oversized contact	
5 masks	9 masks	Metal	Metal
		Passivation	Passivation
		8 masks	10 masks

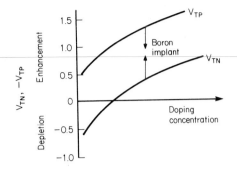

Figure 11-10 Threshold voltage as a function of doping concentration for n-doped poly, showing a 1.0-V difference in ϕ_{MS}.

point is evident from a comparison of HMOS and high-performance CMOS in the same table. Overall, a good rule of thumb is that CMOS increases wafer costs by 20 percent over equivalent single channel versions.

A third issue is the doping polarity of the poly over p-channel devices. Using p-type poly reduces the absolute value of the p-channel threshold voltage, obviating the need for threshold adjust implants. However, it requires a metal strap to short out the diode wherever p-type and n-type poly meet. It has also been found that boron in the poly penetrates into the gate oxide, causing p-channel threshold instability. Using all n-type poly is now more common. However, this creates a 1.0-V difference in ϕ_{MS} between the two channel types, as Fig. 11-10 indicates. To adjust the n-channel threshold upward and p-channel threshold downward such that they end up similar in magnitude, a single-blanket boron implant can be performed. The thresholds will be adjusted in the directions indicated in the figure.

11-6 LATCH-UP

CMOS circuits are susceptible to latch-up because of the presence of a four-layer $pnpn$ structure. During latch-up, the CMOS circuit presents a near short-circuit condition across the power supply. If current flow is not limited by external means, some metal or diffusion current paths will blow open. If current is limited to a safe level, turning the power supply off and then back on will eliminate the condition.

The four-layer structure for the p-tub process comes from the p^+ source and drain, n substrate, p-tub, and n^+ source and drain (see Fig. 11-11). Under normal biasing conditions, all junctions are reverse-biased. However, when one of the source and drain junctions is forward-biased, even momentarily, internal gain amplifies the current until latch-up results. Forward biasing can come from voltage overshoots on an output lead, electrostatic discharge to an input lead, or application of signal levels to leads before power is applied to the circuit. The figure also shows an equivalent circuit to represent the $pnpn$ structure that consists of a vertical npn and a lateral *pnp* bipolar transistor. The vertical *npn* transistor is of fairly high current gain, from 80 to 500, while the lateral *pnp* is of low gain, 0.01 to 1. To analyze the condition for latch-up, break the feedback loop and consider the gain around the loop. $\beta_{npn} = I_Y/I_X$ and $\beta_{pnp} = I_X/I_Y'$. Positive feedback is present if $I_Y/I_Y' > 1$, or

(11-5) $\quad \beta_{npn}\beta_{pnp} > 1$

The two-transistor model can be refined with the addition of the resistances in the current path, as shown in Fig. 11-12. R_S and R_W are the resistances in the substrate and well, respectively. The figure shows that if the emitter of the lateral *pnp* is forward-biased, then the current flow is primarily hole current; whereas if the vertical *npn* is forward-biased, the current flow is predominantly electron current. With such a model in place, one can discuss the effectiveness of the general ways of preventing latch-up, namely by layout techniques and process changes.

Prevention with Layout Techniques

The layout of input and output circuits requires close attention because they are the most likely places where latch-up initiates. The first solution in layout is to increase spacing between the *p*-channel drains and any surrounding *p*-tubs. The purpose is to cut down the lateral gain as well as increase R_1 to increase the latch-up holding voltage. Since minority carriers in the lightly doped base (substrate) have a rather long diffusion length, lateral

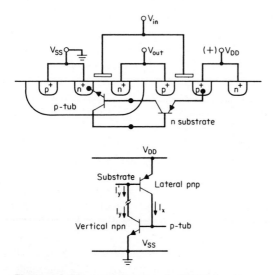

Figure 11-11 *pnpn* Structure in CMOS that is susceptible to latch-up. Its equivalent two-transistor model is also shown.

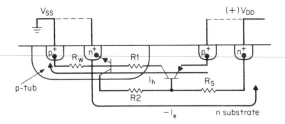

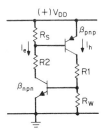

Figure 11-12 Resistances in the current path.

gain falls off slowly with distance. As a result, separations of 75 μm or more are needed for this solution to be effective.

A second layout solution, in fact the most effective, is the use of guard rings. Contact to the tub, in the form of a p^+ diffusion is placed between any n-channel drains and the edge of the p-tub. A similar n^+ guard ring makes contact to the substrate just outside the well, as Fig. 11-13 shows. The presence of the guard ring on the substrate side cuts down the lateral gain significantly. But most importantly, the guard rings shunt R_S and R_W. The latch-up currents I_h and I_e need to develop one V_{be} drop across R_W and R_S, respectively, to initiate latch-up. Specifically, summing currents at nodes 1 and 2, latch-up occurs when

(11-6a) $\quad I_h = V_{be} \left(\dfrac{1}{R_W} + \dfrac{1}{R_{PGR}} + \dfrac{1}{R_S \beta_{npn}} \right)$

(11-6b) $\quad I_e = V_{be} \left(\dfrac{1}{R_S} + \dfrac{1}{R_{NGR}} + \dfrac{1}{R_W \beta_{pnp}} \right)$

The addition of R_{PGR} and R_{NGR} increases the forcing currents I_h and I_e before latch-up will occur. In other words, the guard rings make it that much harder to develop one V_{be} drop to turn on the second transistor in the latch-up pair.

By the same reasoning, a large number of contacts to wells or substrates will help short out stray well or substrate currents and help prevent a large enough voltage from being developed to turn on a junction. Yet another example of the same principle is shown in the input protection circuitry of

Fig. 11-14. The protection diodes are completely surrounded by guard rings. The concept is that when the protection diodes are forward-biased during an electrostatic discharge, the current flow is shunted or collected by the guard ring rather than spreading out to serve as base current for other transistors.

The solutions enumerated thus far are for latch-up currents because of junctions that are forward-biased by external sources. Within the interior of a chip, there are other sources of latch-up currents. One source is a substrate current caused by carrier avalanche multiplication near the drain. This problem is particularly severe for n-channel devices whose gate oxide

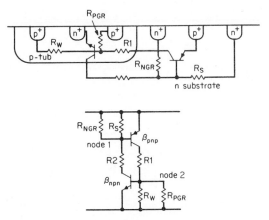

Figure 11-13 Guard rings tied to the power supply and placed near the p-well edge minimize latch-up.

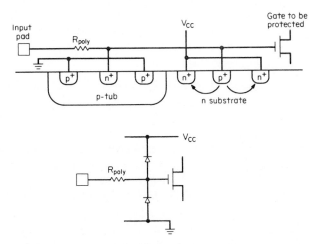

Figure 11-14 Use of guard rings to surround input protection circuitry for latch-up prevention.

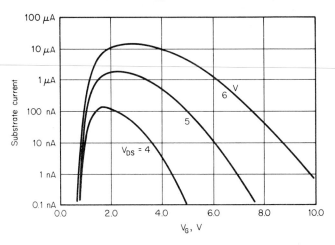

Figure 11-15 Substrate current peaks for a particular value of gate voltage and increase with drain voltage.

thicknesses are scaled down without corresponding reduction in supply voltages. Figure 11-15 shows the nature of this current. It peaks at a rather low value of gate voltage and increases very rapidly with drain voltage. For the same drain voltage, the peak current increases as the oxide thickness is reduced. Another source of latch-up current is parasitic field transistors whose threshold is exceeded. In Fig. 11-11 one can visualize a p-channel field transistor that injects current from the p^+ drain to p-tub. This places the field transistor in parallel with the lateral pnp transistor. A similar situation exists between an n-channel drain and the substrate. In all such cases, the use of a double guard ring (both p^+ and n^+ guard rings) and liberal use of substrate and well contacts are the solution.

Prevention with Process Techniques

Immunity to latch-up can be built into the process itself. Increasing the well junction depth to decrease the vertical β is straightforward. However, the penalty in layout density makes it unpalatable to circuit designers. A very effective solution is the use of insulating substrate such as the silicon-on-sapphire process (see next section) to eliminate completely tub and substrate current flow. Decreasing minority carrier lifetime by gold doping or neutron irradiation[5] has also shown to be effective. However, such a procedure produces leaky junctions and is a difficult process to control. One process that holds a great deal of promise is the use of an epitaxial layer on very low resistivity substrate as shown in Fig. 11-16. The epitaxial layer has to be thick enough to avoid punch-through between well and substrate. The low-

resistivity substrate effectively shorts out any voltage drop caused by substrate currents. A further refinement is to use a buried layer under every
well as shown in Fig 11-17. The buried layer effectively shorts out the well
resistance and kills the vertical gain by heavily doping the base. The epitaxial layer needs to be thin for the well to merge with the buried layer. The
substrate should be lightly doped to sustain high well-to-substrate breakdown voltage.

The testing of susceptibility to latch-up is not well standardized. To test
internal circuits, the part is exercised at full speed at increasingly higher
power supply voltage until the part latches up. The part is most susceptible
at high temperature because bipolar transistor gains are highest and V_{be}
drops are lowest. To test input-output circuits, the leads are brought more
positive than V_{DD} to force current into the lead and then brought more negative than ground (V_{SS}) to force current out of the lead. The part is deemed
latch-up proof if no latch-up occurs after forcing a certain predetermined
current, ranging from 10 to 200 mA.

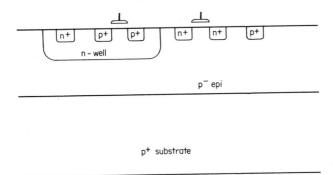

Figure 11-16 Use of an epitaxial layer on low-resistivity substrate to minimize latch-up.

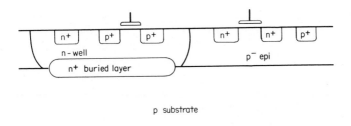

Figure 11-17 Use of buried and epitaxial layers to minimize latch-
up.

11-7 SILICON ON SAPPHIRE

One CMOS process that deserves special discussion is silicon on sapphire (SOS). MOS transistors are fabricated on an insulating substrate, with both p and n channels available for constructing CMOS circuits. Sapphire (Al_2O_3) is a popular choice for the substrate because it has a fairly close match in lattice constant with silicon. Sapphire substrates are translucent and much less fragile than silicon wafers.

The fabrication sequence[6] is shown in Fig. 11-18. A 0.6- to 1.0-μm silicon epitaxial layer is grown on top of the sapphire substrate (step a). An isotropic etch is used to separate the silicon layers into isolated islands (step b). The appropriate p, n^+, and p^+ diffusions are then diffused in (steps c and d). A thick layer of silicon dioxide is deposited, gate areas etched, and 1000 Å of gate oxide grown (step e). Finally, contact holes are etched and

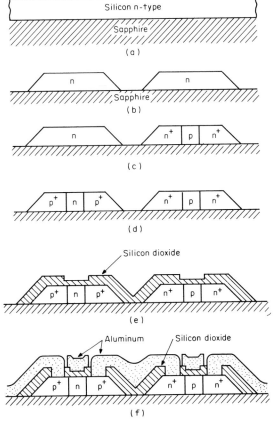

Figure 11-18 Fabrication sequence for silicon on sapphire (SOS) CMOS circuits. (*After Ref. 6.*)

metal deposited and patterned (step *f*). An improvement on the process sequence as shown is to replace the isotropic etch with a field oxidation step similar to LOCOS (see Sec. 7-4). The field oxide would extend all the way to the sapphire substrate.

SOS is built with each transistor having its own body (substrate), which is floating. But the substrate potential remains fixed at one V_{be} above or below the source and does not present a problem.

Because all drain junctions terminate in the insulating substrate, drain capacitance is extremely low. Parasitic capacitance due to interconnection is likewise virtually eliminated. The result is substantial improvement in speed. Counters with operating frequencies of over 100 MHz and memories with 18 ns access time are possible.[7] SOS also occupies less area because of the elimination of guard rings and wells. But the biggest advantage of SOS lies in the absence of CMOS latch-up. This makes SOS a good choice for radiation-hardened CMOS.[8] SOS is particularly immune to transient radiation which causes photo currents to flow across *pn* junctions and to upset the logic in most other processes.

SOS suffers from higher source-to-drain leakage current than the bulk CMOS process. The silicon-to-sapphire interface is not perfect and is the source of most of the leakage. Carrier lifetimes in the epitaxial layer are also often lower than bulk lifetimes, thus contributing to the leakage. The carrier mobilities of SOS transistors are also lower than those in the bulk process, but in recent years the difference has become insignificant. One disadvantage that has not been mitigated is the higher cost of the sapphire substrate, which is three to five times more than an equivalent silicon wafer.

An alternative to SOS is silicon on insulator[9] (SOI). This refers to growing 0.5 to 1.0 μm of SiO_2 on a Si wafer, followed by deposition of 0.5 μm of polysilicon. The polysilicon is then converted to a single crystal by recrystallization techniques such as laser, E-beam annealing, incandescent lamp, or graphite strip heating. The advantages over SOS are the lower cost for silicon wafers and a better quality Si–SiO_2 substrate interface. It is an area under active investigation.

REFERENCES

1. J. Posa, "C-MOS inspires the best chip yet for computer, consumer, and communication applications," *Electronics,* pp. 103–126, Oct. 6, 1981.

2. K. Yu et al., "CMOS static RAM feels at home with ECL speeds," *Electronics,* pp. 160–163, Feb. 10, 1982.

3. E. R. Hnatek, *User's Guidebook to Digital CMOS Integrated Circuits,* Chap. 2, McGraw-Hill, New York, 1981.

4. K. Yu et al., "HMOS-CMOS—A low-power high performance technology," *IEEE J. Solid-State Circ.,* vol. SC-16, no. 5, pp. 454–459, October 1981.

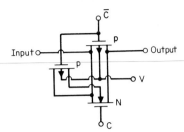

Figure 11-19 Prob. 4.

5. W. Dawes and G. F. Derbenwick, *IEEE Trans. Nuclear Sci.,* vol. NS-23, December 1976.

6. E. R. Hnatek, *A User's Handbook of Semiconductor Memories,* Chap. 2, Wiley, New York, 1977.

7. M. Isobe et al., "An 18 ns CMOS/SOS 4K static RAM," *Proc. ISSCC,* pp. 12–13 and 252, Feb. 18,1981.

8. H. Borkan, "Radiation hardening of CMOS technologies—an overview," *IEEE Trans. Nuclear Sci.,* vol. NS-24, no. 6, pp. 2043–2046, December 1977.

9. VLSI Laboratories, T.I., "Technology and design challenges of MOS VLSI," *IEEE J. Solid-State Circ.,* vol. SC-17, no. 3, pp. 442–448, June 1982.

PROBLEMS

1. Derive Eq. (11-1) by equating the saturation currents through both devices when $V_{out} = V_{transition} = V_{in}$.

2. A CMOS inverter is driving a 1-pF load. It has a leakage current of 100 nA. On log-log paper, plot the static power dissipation, dynamic power dissipation, and total power dissipation as a function of frequency from 1 kHz to 1 MHz. $V_{DD} = 10$ V.

3. Calculate the switching point for a CMOS inverter with $\beta_N = 5$ $\mu A/V^2$, $\beta_P = 1$ $\mu A/V^2$, $V_{TN} = 0.5$ V, $V_{TP} = -0.4$ V, $V_{DD} = 10$ V. Repeat the calculations for $V_{DD} = 5$ V. Sketch the two transfer characteristics.

4. A normal CMOS transmission gate is modified by the addition of a second p-channel device to bias the body of the n-channel device (see Fig. 11-19). Roughly sketch what the n-channel on-resistance of Fig. 11-4 would change to. What would the new composite resistance be as a function of input bias?

12

Computer Circuit Simulation

12-1 NEED FOR CIRCUIT SIMULATION PROGRAMS

The use of circuit simulation programs is indispensable in the design of MOS integrated circuits. It allows designing for proper circuit operation even under worst-case conditions.

As the complexity of integrated circuits increases, it is no longer practical to simulate circuit performance with breadboards built out of single logic gates or MSI (medium-scale integration) ICs. Those discrete packages do not have the same input-output interface characteristics as devices on the final integrated circuit. Furthermore, the problem of large parasitic capacitances and the dubious dependability of many hand-soldered connections make breadboarding a very poor technique for verifying circuit design.

To verify circuit performance test chips containing key circuit blocks could be built. However, there is a considerable time delay in getting the circuit back for evaluation. For this reason, test chips are often used for final verification only *after* circuit design techniques have been developed.

The more appropriate methodology is to use computer simulation. Considerable effort is spent ahead of time to develop accurate models of the devices in a given process. The accuracy of the models has to be balanced against the computational complexity, considering that each device model is utilized thousands of times in each simulation. Once a model is constructed for a typical device, together with its associated set of device parameters, then the worst-case models should also be constructed. Ideally, those worst-case models are generated from actual devices obtained through processing extremes; however, in the absence of such devices, reasonable extrapolation of the device parameters is appropriate. Equal attention is paid not just to modeling active devices, but also to the parasitic capacitances and parasitic resistances.

With good active and parasitic device models, and with an efficient circuit analysis program, circuit performance can be optimized.

12-2 ASPEC CIRCUIT SIMULATION PROGRAM

Most circuit simulation programs are similar in general format because the required input information and desired outputs are similar. So, rather than discuss circuit simulation using generalized concepts, a specific circuit simulator will be discussed in this chapter. The program ASPEC (Advanced Simulation Program for Electronic Circuits) will be used as a specific example for performing transient and ac analysis. Because of page limitations, only a subset of the ASPEC program can be discussed. Two sample runs will be shown, followed by a condensed version of the user's manual in the appendix following this chapter.

Special Note: ASPEC is a commercially available program offered exclusively by Information Systems Design, a division of Control Data Corporation located at 2500 Mission College Boulevard, Santa Clara, California 95050. The author is grateful for the permission received to reproduce portions of the ASPEC user's manual. The choice of ASPEC for discussion does not constitute an endorsement by the author or the publisher.

ASPEC has more capabilities than are represented in this chapter. Notable features not covered are (1) the ability to perform dc and transfer function analysis; (2) the availability of elements such as inductors, transconductance, voltage-controlled switches, batteries, coupled inductors, Schottky diodes, bipolar junction transistors, junction FETs, sinusoidal and exponential sources; and (3) the definition of groups of sources, elements, and devices as a single functional block called "macro" for repetitive use in a circuit.

An explanation of how to use ASPEC will be given below, together with two sample runs, one on the transient analysis of a two-input CMOS NAND gate and the other on an ac analysis of a CMOS operational amplifier. To understand fully the workings of the program, one must also read and understand the condensed user's manual. Most people find it more enlightening to go through the example first and then read the appendix for the specific format rules.

ASPEC requires an input card deck (or input data file) that contains information describing the circuit to be simulated. All the nodes have to be identified, as well as all circuit elements. External excitations such as power supplies and signal sources have to be described too. Parasitic elements can be represented as additional circuit elements. Internal to the program are models for the circuit elements, although specific parameters for those models can be altered when needed. Once the simulation is completed, the

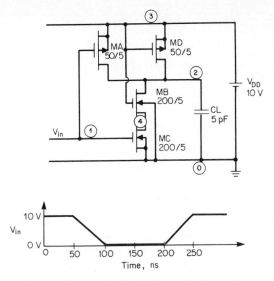

Figure 12-1 Two-input CMOS NAND gate as an example for a transient analysis run.

manner in which results are presented is specified. Output response can be in the form of a tabular printout or in the form of a plot generated on the printer. Device information such as current flow and power dissipation are also available. All the above information is entered with specific key words to indicate the type of data. This allows the data to be entered in any order, with the exception of the first and last card.

12-3 TRANSIENT ANALYSIS SAMPLE RUN

The response of a two-input CMOS NAND gate, shown in Fig. 12-1, will be determined from a sample transient analysis ASPEC run. The first step is to label all nodes with node numbers. The only restriction on the node number values is that the datum or ground node be node 0. Next, the circuit elements are given names, such as MA, MB, CL, or VDD. Then the input file can be built.

Input File

The input file for the run is shown in Fig. 12-2. The numbers preceding each line are for easy reference for the following discussion and should not be included in the actual input. The very first line or card has to be a title or heading for the run. This heading is required and is printed at the top of every output page.

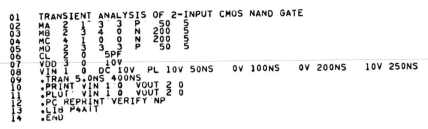

Figure 12-2 Input file for a transient analysis run.

The next four lines describe the four MOS transistors used in the circuit. The device name is the first parameter in each line, and the names have to begin with the letter M to indicate a MOS transistor. The next four parameters in each line are node numbers of the drain, gate, source, and body, respectively. Recall that the source and drain of MOS transistors are interchangeable; thus ASPEC is set up to accept either lead assignment. The next parameter is a name for the device model, which in this case is either N or P. The characteristics of the devices N and P are either internal to the program or specified elsewhere in the input file. The next two numbers describe the W and L of the device in micrometers.

Line 6 describes the capacitor. The device name begins with C to indicate that this is a capacitor. The next two parameters show that it is connected between node 2 and ground. The final parameter is the capacitance value.

Line 7 is for the voltage source. The fact that the element name begins with V indicates a voltage source. This source is connected to node 3 and ground, with node 3 being the positive node. The single voltage value that follows indicates a dc voltage source. Current sources would begin with an I.

Line 8 shows another voltage source, connected between node 1 and 0. It has an initial dc value of 10 V, followed by a *p*iecewise *l*inear waveform described by voltage-time breakpoint pairs. The breakpoint pairs describe the input waveform shown in Fig. 12-1. After the last time breakpoint, a constant voltage value is assumed.

All the lines described thus far are for circuit components since they all begin with a letter. The next set of lines are instructions and are identified by beginning with a period.

Line 9 indicates a transient analysis run. One, and only one, such line is needed in every run to specify the type of analysis. That line requests a transient analysis for 400 ns, with a printout every 5-ns time step. The time step is used only for printout; ASPEC itself uses considerably smaller time steps for its calculations.

Line 10 requests a printout of two voltages, one between nodes 1 and 0, and the other between 2 and 0. The names of the output quantities, VIN and VOUT, are arbitrary and are for the user's own benefit. However, they do have to begin with V to indicate printout of voltage. Printout of current and power through an element is also possible by specifying device names instead of node numbers. In addition, current through a MOSFET is available by specifying the first two characters of the output name as ID, IG, IS, or IB for drain, gate, source, or bulk current, respectively.

Line 11 requests a printout on the printer in graphical form. The request formats are the same as for a .PRINT line. The .PRINT and .PLOT lines need not request for identical outputs as the example does.

Line 12 is the print control line (.PC) which controls the amount and format of the program printout. The options after .PC for this sample run are REPRINT, VERIFY, and NP. REPRINT provides a printout of all the device elements, excitation sources, instruction cards, and device model data. VERIFY causes a printout of a node connection table, analysis iteration control parameters, and program storage usage. NP requests printout to fit a narrow paper.

Line 13 allows calling device model parameters from a preestablished library arbitrarily called P4XTT and which contains the device models P and N. This does not preclude the use of a .MODEL card to alter specific device model parameters, if necessary.

The last line in the input file has to be an .END card to signify end of run. Though not shown in the sample run, an .ALTER card can be used before the .END card to initiate another run. The altered cards with their new values then follow the .ALTER card. The .ALTER card may be used several times with all the changes valid for the subsequent run unless specifically changed back to the original.

Program Output

The output from the sample run is reproduced in Fig. 12-3. First is a printout of the input card file to verify proper reading in of the input file. Figure 12-3*b* is the result of the .PC REPRINT option. All the instructions, sources, circuit elements, and device models are listed. Figure 12-3*c* is the result of the VERIFY option on the same print control card. It provides a node table to verify circuit connections. This is followed by analysis iteration control parameters and program storage utilization. Figure 12-3*d* is a tabulated printout of outputs as specified by the .PRINT card, and Fig. 12-3*e* gives the output waveforms as "plotted" on the printer.

12-4   AC ANALYSIS SAMPLE RUN

A CMOS operational amplifier shown in Fig. 12-4 is used as an example of ac analysis with ASPEC. The numbering of nodes and naming of circuit

```
*********************** A S P E C ***************************
TRANSIENT ANALYSIS OF 2-INPUT CMOS NAND GATE
                        LAST UPDATED ON 11/9 1979
**************************************** (M) BI ******************************
                        DATE 09/22/82   12:18:51
```

```
MA  2  1  3  3  P   50   5
MB  2  3  4  0  N  200   5
MC  4  1  0  0  N  200   5
MD  2  3  3  3  P   50   5
CL  2  0   5PF
VDD 3  0   10V
VIN 1  0   DC 10V  PL 10V 50NS   0V 100NS   0V 200NS   10V 250NS
.TRAN 5.0NS 400NS
.PRINT VIN 1 0   VOUT 2 0
.PLOT  VIN 1 0   VOUT 2 0
.PC REPRINT VERIFY NP
.LIB P4XTT
         ,0.,0.,0.,0./
.END
```

Figure 12-3 Program output for transient analysis run. (*a*) Input file.


```
** A S P E C **                    DATE 09/22/82   12:18:51
TRANSIENT ANALYSIS OF 2-INPUT CMOS NAND GATE


.TEMP    25.0

.TRAN    5.000D-09 4.0000D-07


.PRINT          VIN        * 1.0D+00  1    0
                VOUT       * 1.0D+00  2    0

.PLOT           VIN        * 1.0D+00  1    0
                VOUT       * 1.0D+00  2    0
```

V AND I SOURCES -

NAME	NODES		TYPE	VALUES			
VDD	3	0	DC	1.000D+01			
VIN	1	0	DC	1.000D+01			
			PL	1.000D+01	5.000D-08		
				0.0	1.000D-07	0.0	2.000D-07
				1.000D+01	2.500D-07		

LINEAR ELEMENTS -

NAME	NODES		VALUE	TC (RB)	Z (TCV)	L
CL	2	0	5.0000D-12	0.0		

MOS FIELD EFFECT TRANSISTORS -

NAME	D	G	S	B	MODEL VDS	WIDTH VGS	LENGTH VSB	(W/L)EFF WEFF	LEFF
MA	2	1	3	3	P 1.00	50.00 2.70	5.00 0.0	16.79 47.00	2.80
MB	2	3	4	0	N 1.00	200.00 2.46	5.00 0.0	65.90 197.70	3.00
MC	4	1	0	0	N 1.00	200.00 2.46	5.00 0.0	65.90 197.70	3.00
MD	2	3	3	3	P 1.00	50.00 2.70	5.00 0.0	16.79 47.00	2.80

Figure 12-3   (**continued**)

```
DEVICE MODELS -
.MODEL   N          NMOS    X0   = 4.00D-02     DFD = 2.00D-01
                            DFL  = 2.50D-01     COVW= 2.11D-16
                            CBO  = 1.30D+15     CKJA= 9.45D-17
                            QSS0=-2.00D+11      FSS = 2.000+11
                            QSSP= 2.00D+10      PHMS=-5.50D-01
                            CJPG= 1.48D-16      VBPG= 8.70D-01
                            EXPG= 2.70D-01      ENA = 5.00D+19
                            ENS = 1.70D+16      XI  = 3.30D-01
                            PHFS=-1.00D+00      U0  = 8.00D+02
                            ES0 = 4.00D+04      EC  = 3.90D+04
                            PTP = 2.00D+00      ALPH= 2.00D-01
                            BETA= 6.00D-01      EXPO= 1.11D-01
                            AEXP= 5.00D-01      BEXP= 9.00D-01
                            ZDEL=-2.30D+00      NF  = 5.00D+15
                            FLD = 2.00D+00      DELL=-1.50D+00
                            OHMS= 3.00D-02      KF  = 0.0
                            AF  = 1.00D+00

.MODEL   DJN        D       IS  = 2.06D-18     N   = 1.00D+00
                            TT  = 0.0           CJA = 9.45D-17
                            CJP = 1.48D-16      PHI = 8.70D-01
                            EXA = 5.00D-01      EXP = 2.70D-01
                            CFX = 8.63D-16      CTC = 0.0
                            EG  = 1.11D+00      EX  = 3.00D+00
                            XOI = 1.05D+00      XOM = 1.40D+00
                            ISP = 0.0           RS  = 0.0
                            KF  = 0.0           AF  = 1.00D+00
                            TRS = 0.0           VRB = 1.00D+03
                            RZ  = 1.00D+01      TCV = 0.0

.MODEL   P          PMOS    X0   = 4.00D-02     DFD = 1.00D-01
                            DFL  = 3.50D-01     COVW= 1.90D-16
                            CBO  = 1.20D+16     CKJA= 2.10D-16
                            QSS0= 3.00D+10      FSS = 2.000+11
                            QSSP=-8.00D+10      PHMS= 0.0
                            CJPG= 5.30D-16      VBPG= 9.30D-01
                            EXPG= 2.70D-01      ENA = 5.00D+19
                            ENS = 1.20D+16      XI  = 1.00D+09
                            PHFS= 1.00D+00      U0  = 3.60D+02
                            ES0 = 2.80D+04      EC  = 2.00D+05
                            PTP = 8.00D-01      ALPH= 2.00D-01
                            BETA= 6.00D-01      EXPO= 3.00D-01
                            AEXP= 2.60D-01      BEXP= 2.00D-01
                            ZDEL=-3.00D+00      NF  =-1.30D+15
                            FLD = 2.60D+00      DELL=-1.50D+00
                            OHMS= 7.00D-02      KF  = 0.0
                            AF  = 1.00D+00

.MODEL   DJP        D       IS  = 2.06D-18     N   = 1.00D+00
                            TT  = 0.0           CJA = 2.10D-16
                            CJP = 5.30D-16      PHI = 9.30D-01
                            EXA = 5.00D-01      EXP = 2.70D-01
                            CFX = 8.63D-16      CTC = 0.0
                            EG  = 1.11D+00      EX  = 3.00D+00
                            XOI = 1.05D+00      XOM = 1.40D+00
                            ISP = 0.0           RS  = 0.0
                            KF  = 0.0           AF  = 1.00D+00
                            TRS = 0.0           VRB = 1.00D+03
                            RZ  = 1.00D+01      TCV = 0.0
```

Figure 12-3   (continued)   (*b*) .PC REPRINT option.

elements and voltage sources follow the same convention as that for transient analysis.

Input File

The input file is shown in Fig. 12-5. Lines 2 through 10 describe all the transistors, their connections, device type and size. Similarly, capacitors C1 and CL are defined in lines 11 and 12. The power supplies VBB and VCC are defined in lines 13 and 14. The noninverting input VIN+ has a dc voltage of 0 V (ground) and an ac voltage of 1 V. The inverting input VIN− is grounded.

```
** A S P E C **                          DATE 09/22/82    12:18:51

TRANSIENT ANALYSIS OF 2-INPUT CMOS NAND GATE


NODE CONNECTIONS -

    0       VDD       VIN       CL        MB        MC

    1       VIN       MA        MC

    2       CL        MA        MB        MD

    3       VDD       MA        MB        MD

    4       MB        MC


.CONTROL
    ABSV=1.00D-03        RELV=1.00D-01      ABSI=1.00D-09      RELI=5.00D-02
    RMAX=1.00D+00        RMIN=1.00D-03      IMAX= 5            IMIN= 3
    SCUT= 20.            VLIM=0.20          VCHG=0.80          DCSTEP=1.00D+00


PROGRAM STORAGE UTILIZATION -

PARAMETER                USAGE    LIMIT

TOTAL NODES                5       800
MATRIX STORAGE             7      5000
TOTAL SOURCES              2        30
TOTAL ELEMENTS             1       400
TOTAL DEVICES              4      2000
DEVICE MODELS              4        40
TIME PRINT INTERVALS       1        20
TOTAL OUTPUTS              2       800
.PRINT & .PLOT CARDS       2       100
TEMPERATURE VALUES         1        10
.DCVOLT VOLTAGES           0       700
FREQUENCY INTERVALS        0         5
TOTAL .PARAM NAMES         0       200
TOTAL .PARAM USES          0       400
.MACRO DEFINITIONS         0        50
.MACRO EXPANSIONS          0       200
```

Figure 12-3 (continued) (*c*) .PC VERIFY option.

At this juncture, it must be pointed out that ASPEC first solves for dc operating point, then extracts linearized ac parameters around that operating point. AC analysis then consists of obtaining the response of the linearized equivalent circuit to any ac input. The magnitude of the ac input then has no effect on the dc operating point. For example, large ac signals would not result in clipping in the analysis. To simulate large signal behavior, the transient analysis should be used, with all input signals in actual time varying, sinusoidal waveforms.

Line 17 is an instruction card to perform frequency response analysis. The analysis extends between 100 Hz to 10 MHz with 10 points per decade printed or plotted on a logarithmic scale. Other options aside from DEC are OCT and LIN, indicating logarithmic (in octaves) and linear frequency printouts, respectively.

The .PRINT and .PLOT cards are similar to that for transient analysis except that the type or component of the ac output must be specified. The types are MAG, PHAS, DB, REAL, and IMAG. They correspond to magnitude, phase, 20*log(magnitude), real part, and imaginary part, respectively, of the requested output.

TRANSIENT ANALYSIS OF 2-INPUT CMOS NAND GATE
TEMPERATURE = 25.0
TRANSIENT ANALYSIS -

TIME	VIN * 1.0D+00 1 0	VOUT * 1.0D+00 2 0
0.0	10.00000	0.00000
5.0000D-09	10.00000	-0.00000
1.0000D-08	10.00000	-0.00000
1.5000D-08	10.00000	0.00000
2.0000D-08	10.00000	0.00000
2.5000D-08	10.00000	0.00000
3.0000D-08	10.00000	0.00000
3.5000D-08	10.00000	0.00000
4.0000D-08	10.00000	0.00000
4.5000D-08	10.00000	0.00000
5.0000D-08	10.00000	0.00000
5.5000D-08	9.00000	-0.00271
6.0000D-08	8.00000	0.00714
6.5000D-08	7.00000	0.03221
7.0000D-08	6.00000	0.07319
7.5000D-08	5.00000	0.13485
8.0000D-08	4.00000	0.23117
8.5000D-08	3.00000	0.40976
9.0000D-08	2.00000	1.17328
9.5000D-08	1.00000	3.38136
1.0000D-07	0.00000	5.94442
1.0500D-07	0.0	8.03631
1.1000D-07	0.0	9.20271
1.1500D-07	0.0	9.76958
1.2000D-07	0.0	9.95207
1.2500D-07	0.0	9.99770
1.3000D-07	0.0	10.00570
1.3500D-07	0.0	10.00507
1.4000D-07	0.0	10.00355
1.4500D-07	0.0	10.00171
1.5000D-07	0.0	9.99968
1.5500D-07	0.0	9.99828
1.6000D-07	0.0	9.99802
1.6500D-07	0.0	9.99844
1.7000D-07	0.0	9.99899
1.7500D-07	0.0	9.99939
1.8000D-07	0.0	9.99958
1.8500D-07	0.0	9.99965
1.9000D-07	0.0	9.99967
1.9500D-07	0.0	9.99969
2.0000D-07	0.0	9.99971
2.0500D-07	1.00000	10.00783
2.1000D-07	2.00000	9.12526
2.1500D-07	3.00000	6.13950
2.2000D-07	4.00000	1.43267
2.2500D-07	5.00000	0.10262
2.3000D-07	6.00000	0.08037
2.3500D-07	7.00000	0.04693
2.4000D-07	8.00000	0.01662
2.4500D-07	9.00000	0.00310
2.5000D-07	10.00000	0.00192
2.5500D-07	10.00000	-0.00086
2.6000D-07	10.00000	-0.00020
2.6500D-07	10.00000	0.00000
2.7000D-07	10.00000	0.00001
2.7500D-07	10.00000	0.00000
2.8000D-07	10.00000	-0.00000
2.8500D-07	10.00000	-0.00000
2.9000D-07	10.00000	-0.00000
2.9500D-07	10.00000	0.00000
3.0000D-07	10.00000	-0.00000
3.0500D-07	10.00000	-0.00000
3.1000D-07	10.00000	-0.00000
3.1500D-07	10.00000	-0.00000
3.2000D-07	10.00000	-0.00000
3.2500D-07	10.00000	-0.00000
3.3000D-07	10.00000	-0.00000
3.3500D-07	10.00000	-0.00000
3.4000D-07	10.00000	-0.00000
3.4500D-07	10.00000	-0.00000
3.5000D-07	10.00000	-0.00000
3.5500D-07	10.00000	-0.00000
3.6000D-07	10.00000	-0.00000
3.6500D-07	10.00000	-0.00000
3.7000D-07	10.00000	-0.00000
3.7500D-07	10.00000	-0.00000
3.8000D-07	10.00000	-0.00000
3.8500D-07	10.00000	-0.00000
3.9000D-07	10.00000	-0.00000
3.9500D-07	10.00000	-0.00000
4.0000D-07	10.00000	-0.00000

Figure 12-3 **(continued)** (*d*) .PRINT output.

TRANSIENT ANALYSIS OF 2-INPUT CMOS NAND GATE
TEMPERATURE = 25.0
TRANSIENT ANALYSIS -
0) VIN   * 1.00+00 1   0
1) VOUT   * 1.00+00 2   0

```
      TIME
            -1.50        1.50        4.50        7.50       10.50        13.50
           I+++++++++I+++++++++I+++++++++I+++++++++I+++++++++I
0.0        .      1         .         .         .    0    .
5.0000D-09 .      1         .         .         .    0    .
1.0000D-08 .      1         .         .         .    0    .
1.5000D-08 .      1         .         .         .    0    .
2.0000D-08 .      1         .         .         .    0    .
2.5000D-08 .      1         .         .         .    0    .
3.0000D-08 .      1         .         .         .    0    .
3.5000D-08 .      1         .         .         .    0    .
4.0000D-08 .      1         .         .         .    0    .
4.5000D-08 .      1         .         .         .    0    .
5.0000D-08 .      1         .         .         .    0    .
5.5000D-08 .      1         .         .       0   0 .
6.0000D-08 .      1         .         .       0 .
6.5000D-08 .      1         .         .     0 .
7.0000D-08 .      1         .         .   0 .
7.5000D-08 .      1         .        0 .
8.0000D-08 .     1         .      0 .
8.5000D-08 .    1         .
9.0000D-08 .       1. 0  0 .
9.5000D-08 .    0      1  .
1.0000D-07 .  0         .         .         . 1 .
1.0500D-07 .  0         .         .         .  1 .
1.1000D-07 .  0         .         .         .    1 .
1.1500D-07 .  0         .         .         .     1 .
1.2000D-07 .  0         .         .         .     1 .
1.2500D-07 .  0         .         .         .     1 .
1.3000D-07 .  0         .         .         .     1 .
1.3500D-07 .  0         .         .         .     1 .
1.4000D-07 .  0         .         .         .     1 .
1.4500D-07 .  0         .         .         .     1 .
1.5000D-07 .  0         .         .         .     1 .
1.5500D-07 .  0         .         .         .     1 .
1.6000D-07 .  0         .         .         .     1 .
1.6500D-07 .  0         .         .         .     1 .
1.7000D-07 .  0         .         .         .     1 .
1.7500D-07 .  0         .         .         .     1 .
1.8000D-07 .  0         .         .         .     1 .
1.8500D-07 .  0         .         .         .     1 .
1.9000D-07 .  0         .         .         .     1 .
1.9500D-07 .  0         .         .         .     1 .
2.0000D-07 .  0         .         .         .    1 .
2.0500D-07 .      0     .         .         . 1 .
2.1000D-07 .        0   .         .         1 .
2.1500D-07 .         0  .       1 .
2.2000D-07 .      1     0.
2.2500D-07 .  1         .      0 .
2.3000D-07 .  1         .        0 .
2.3500D-07 .  1         .          0 .
2.4000D-07 .  1         .           . 0 .
2.4500D-07 .  1         .         .    0 .
2.5000D-07 .  1         .         .    0 .
2.5500D-07 .  1         .         .    0 .
2.6000D-07 .  1         .         .    0 .
2.6500D-07 .  1         .         .    0 .
2.7000D-07 .  1         .         .    0 .
2.7500D-07 .  1         .         .    0 .
2.8000D-07 .  1         .         .    0 .
2.8500D-07 .  1         .         .    0 .
2.9000D-07 .  1         .         .    0 .
2.9500D-07 .  1         .         .    0 .
3.0000D-07 .  1         .         .    0 .
3.0500D-07 .  1         .         .    0 .
3.1000D-07 .  1         .         .    0 .
3.1500D-07 .  1         .         .    0 .
3.2000D-07 .  1         .         .    0 .
3.2500D-07 .  1         .         .    0 .
3.3000D-07 .  1         .         .    0 .
3.3500D-07 .  1         .         .    0 .
3.4000D-07 .  1         .         .    0 .
3.4500D-07 .  1         .         .    0 .
3.5000D-07 .  1         .         .    0 .
3.5500D-07 .  1         .         .    0 .
3.6000D-07 .  1         .         .    0 .
3.6500D-07 .  1         .         .    0 .
3.7000D-07 .  1         .         .    0 .
3.7500D-07 .  1         .         .    0 .
3.8000D-07 .  1         .         .    0 .
3.8500D-07 .  1         .         .    0 .
3.9000D-07 .  1         .         .    0 .
3.9500D-07 .  1         .         .    0 .
4.0000D-07 .  1         .         .    0 .
           I+++++++++I+++++++++I+++++++++I+++++++++I+++++++++I
```

- END OF RUN -

Figure 12-3 (continued) (e) .PLOT output.

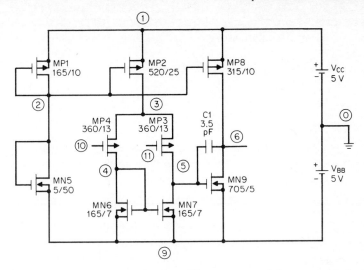

Figure 12-4   CMOS operational amplifier as an example for an ac analysis run.

Line 21 calls device model parameters from a preestablished library called P4XTT. Line 22 is a print control card. In addition to the REPRINT, VERIFY, and NP options already described, ACVER prints out the linearized equivalent circuit parameters used in the analysis.

Line 23 allows control over the analysis iteration scheme. The list of control values defines the conditions necessary before the program halts the iteration and moves on to the next time step. The example requests that the program keep iterating until successive iterations produce changes of no

```
01   AC ANALYSIS OF CMOS OPERATIONAL AMPLIFIER
02   MP1    2   2   1   1   P   165  10
03   MP2    3   2   1   1   P   520  25
04   MP3    5  11   3   1   P   360  13
05   MP4    4  10   3   1   P   360  13
06   MP8    6   2   1   1   P   315  10
07   MN5    2   2   9   9   N     5  50
08   MN6    4   4   9   9   N   165   7
09   MN7    5   4   9   9   N   165   7
10   MN9    6   5   9   9   N   350   7
11   C1     6   5   3.5PF
12   CL     6   0   5.0PF
13   VBB    9   0  -5.0V
14   VCC    1   0   5.0V
15   VIN+  11   0   DC  0V  AC  1V
16   VIN-  10   0   DC  0V
17   .FREQ DEC  10  100  10MEG
18   .PRINT DB  VOUT  6  0   V3  3  0   V4  4  0   V5  5  0
19   .PLOT  DB  VOUT  6  0
20   .PLOT  PHAS VOUT 6  0
21   .LIB P4XTT DATA
22   .PC REPRINT VERIFY ACVER NP
23   .CONTROL ABSI=.1NA ABSV= 1E-03V RELV=1E-03 RELI=1E-03 RMAX=.002
24   .END
```

Figure 12-5   Input file for an ac analysis run.

more than 0.1 nA, 1 mV, 0.1 percent change in voltage, and 0.1 percent change in current, but in any case never less than 500 calculation points per print step.

The input file ends with an .END card.

Program Output

The program output is shown in Fig. 12-6. A listing of the input file is in Fig. 12-6a. Then a tabulation of all instruction, element, and source cards is shown in Fig. 12-6b as a result of the PC reprint option. All transistor parameters and device models are listed too. Figure 12-6c gives a node connection, iteration control, and program storage utilization tables. Figure 12-6d gives the dc operating points, currents through all transistors, and MOSFET operating points. Figure 12-6e provides the linearized ac parameters due to the ACVER print control option. Figure 12-6f is the printout of the voltages at four different nodes. The unit is in volts expressed in decibels. Since the input is at 1-V ac, the numbers are numerically equal to the gain from input to those nodes. Finally, Fig. 12-6g and Fig. 12-6h are the plots of the magnitude and phase of the output voltage.

Some of the conclusions from this analysis run available from Fig. 12-6g and 12-6h are that the op-amp has a maximum open loop gain of 66 dB, a

(*Text continues on page 299.*)

```
*********************   A  S  P  E  C   ************************
AC ANALYSIS OF CMOS OPERATIONAL AMPLIFIER
                         LAST UPDATED ON  11/9 1979
***********************************  (M)  BI  ****************************
                    DATE 09/23/82    12:30:08

MP1   2  2  1  1   P    165 10
MP2   3  2  1  1   P    520 25
MP3   5 11  3  1   P    360 13
MP4   4 10  3  1   P    360 13
MP8   6  2  1  1   P    315 10
MN5   2  2  9  9   N      5 50
MN6   4  4  9  9   N    165  7
MN7   5  4  9  9   N    165  7
MN9   6  5  9  9   N    350  7
C1    6  5   3.5PF
CL    6  0   5.0PF
VBB   9  0  -5.0V
VCC   1  0   5.0V
VIN+ 11  0 DC  0V  AC 1V
VIN- 10  0 DC  0V
.FREQ DEC 10 100 10MEG
.PRINT DB VOUT 6 0  V3 3 0  V4 4 0  V5 5 0
.PLOT DB  VOUT 6 0
.PLOT PHAS VOUT 6 0
.LIB P4XTT DATA
          ,0,,0,,0,,0,/
.PC REPRINT VERIFY ACVER NP
.CONTROL ABSI=.1NA ABSV= 1E-03V RELV=1E-03 RELI=1E-03 RMAX=.002
.END
```

Figure 12-6   Program output for ac analysis run. (*a*) Input file.

```
** A S P E C **                    DATE 09/23/82   12:30:08
AC ANALYSIS OF CMOS OPERATIONAL AMPLIFIER
```

.TEMP   25.0

.FREQ   DEC   10   1.000D+02   1.000D+07

.PRINT	DB	VOUT	* 1.0D+00	6	0
	DB	V3	* 1.0D+00	3	0
	DB	V4	* 1.0D+00	4	0
	DB	V5	* 1.0D+00	5	0
.PLOT	DB	VOUT	* 1.0D+00	6	0
.PLOT	PHAS	VOUT	* 1.0D+00	6	0

V AND I SOURCES -

NAME	NODES		TYPE	VALUES
V88	9	0	DC	-5.000D+00
VCC	1	0	DC	5.000D+00
VIN+	11	0	DC	0.0
			AC	1.000D+00
VIN-	10	0	DC	0.0

LINEAR ELEMENTS -

NAME	NODES		VALUE	TC (RB)	Z (TCV)	L
C1	6	5	3.5000D-12	0.0		
CL	6	0	5.0000D-12	0.0		

MOS FIELD EFFECT TRANSISTORS -

NAME	D	G	S	B	MODEL VDS	WIDTH VGS	LENGTH VSB	(W/L)EFF WEFF	LEFF
MP1	2	2	1	1	P 1.00	165.00 2.70	10.00 0.0	20.77 162.00	7.80
MP2	3	2	1	1	P 1.00	520.00 2.70	25.00 0.0	22.68 517.00	22.80
MP3	5	11	3	1	P 1.00	360.00 2.70	13.00 0.0	33.06 357.00	10.80
MP4	4	10	3	1	P 1.00	360.00 2.70	13.00 0.0	33.06 357.00	10.80
MP8	6	2	1	1	P 1.00	315.00 2.70	10.00 0.0	40.00 312.00	7.80
MN5	2	2	9	9	N 1.00	5.00 2.46	50.00 0.0	0.06 2.70	48.00
MN6	4	4	9	9	N 1.00	165.00 2.46	7.00 0.0	32.54 162.70	5.00
MN7	5	4	9	9	N 1.00	165.00 2.46	7.00 0.0	32.54 162.70	5.00
MN9	6	5	9	9	N 1.00	350.00 2.46	7.00 0.0	69.54 347.70	5.00

Figure 12-6   (continued)

```
DEVICE MODELS -
.MODEL   N            NMOS    XO  = 4.00D-02    DFD = 2.00D-01
                              DFL = 2.50D-01    COVW= 2.11D-16
                              CBO = 1.30D+15    CKJA= 9.45D-17
                              QSSO=-2.00D+11    FSS = 2.00D+11
                              QSSP= 2.00D+10    PHMS=-5.50D-01
                              CJPG= 1.48D-16    VBPG= 8.70D-01
                              EXPG= 2.70D-01    ENA = 5.00D+19
                              ENS = 1.70D+16    XI  = 3.30D-01
                              PHFS=-1.00D+00    UO  = 8.00D+02
                              ESO = 4.00D+04    EC  = 3.90D+04
                              PTP = 2.00D+00    ALPH= 2.00D-01
                              BETA= 6.00D-01    EXPO= 1.11D-01
                              AEXP= 5.00D-01    BEXP= 9.00D-01
                              ZDEL=-2.30D+00    VF  = 5.00D+15
                              FLD = 2.00D+00    DELL=-1.50D+00
                              OHMS= 3.00D-02    KF  = 0.0
                              AF  = 1.00D+00

.MODEL   DJN          D       IS  = 2.06D-18    N   = 1.00D+00
                              TT  = 0.0         CJA = 9.45D-17
                              CJP = 1.48D-16    PHI = 8.70D-01
                              EXA = 5.00D-01    EXP = 2.70D-01
                              CFX = 8.63D-16    CTC = 0.0
                              EG  = 1.11D+00    EX  = 3.00D+00
                              XOI = 1.05D+00    XOM = 1.40D+00
                              ISP = 0.0         RS  = 0.0
                              KF  = 0.0         AF  = 1.00D+00
                              TRS = 0.0         VRB = 1.00D+03
                              RZ  = 1.00D+01    TCV = 0.0

.MODEL   P            PMOS    XO  = 4.00D-02    DFD = 1.00D-01
                              DFL = 3.50D-01    COVW= 1.90D-16
                              CBO = 1.20D+16    CKJA= 2.10D-16
                              QSSO= 3.00D+10    FSS = 2.00D+11
                              QSSP=-8.00D+10    PHMS= 0.0
                              CJPG= 5.30D-16    VBPG= 9.30D-01
                              EXPG= 2.70D-01    ENA = 5.00D+19
                              ENS = 1.20D+16    XI  = 1.00D+09
                              PHFS= 1.00D+00    JO  = 3.60D+02
                              ESO = 2.80D+04    EC  = 2.00D+05
                              PTP = 8.00D-01    ALPH= 2.00D-01
                              BETA= 6.00D-01    EXPO= 3.00D-01
                              AEXP= 2.60D-01    BEXP= 2.00D-01
                              ZDEL=-3.00D+00    VF  =-1.30D+15
                              FLD = 2.60D+00    DELL=-1.50D+00
                              OHMS= 7.00D-02    KF  = 0.0
                              AF  = 1.00D+00

.MODEL   DJP          D       IS  = 2.06D-18    N   = 1.00D+00
                              TT  = 0.0         CJA = 2.10D-16
                              CJP = 5.30D-16    PHI = 9.30D-01
                              EXA = 5.00D-01    EXP = 2.70D-01
                              CFX = 8.63D-16    CTC = 0.0
                              EG  = 1.11D+00    EX  = 3.00D+00
                              XOI = 1.05D+00    XOM = 1.40D+00
                              ISP = 0.0         RS  = 0.0
                              KF  = 0.0         AF  = 1.00D+00
                              TRS = 0.0         VRB = 1.00D+03
                              RZ  = 1.00D+01    TCV = 0.0
```

Figure 12-6 (continued) (*b*) .PC REPRINT option.

```
** A S P E C **                    DATE 09/23/82    12:30:09
AC ANALYSIS OF CMOS OPERATIONAL AMPLIFIER


NODE CONNECTIONS -

  0      VBB      VCC      VIN+     VIN-     CL

  1      VCC      MP1      MP2      MP3      MP4      MP8

  2      MP1      MP2      MP8      MN5

  3      MP2      MP3      MP4

  4      MP4      MN6      MN7

  5      C1       MP3      MN7      MN9

  6      C1       CL       MP8      MN9

  9      VBB      MN5      MN6      MN7      MN9

 10      VIN-     MP4

 11      VIN+     MP3

.CONTROL
     ABSV=1.00D-03      RELV=1.00D-03      ABSI=1.00D-10      RELI=1.00D-03
     RMAX=2.00D-03      RMIN=1.00D-03      IMAX= 5            IMIN= 3
     SCUT= 20.          VLIM=0.20          VCHG=0.80          DCSTEP=1.00D+00

PROGRAM STORAGE UTILIZATION -

PARAMETER              USAGE    LIMIT

TOTAL NODES              10      800
MATRIX STORAGE           23     5000
TOTAL SOURCES             4       30
TOTAL ELEMENTS            2      400
TOTAL DEVICES             9     2000
DEVICE MODELS             4       40
TIME PRINT INTERVALS      0       20
TOTAL OUTPUTS             5      800
.PRINT & .PLOT CARDS      3      100
TEMPERATURE VALUES        1       10
.DCVOLT VOLTAGES          0      700
FREQUENCY INTERVALS       1        5
TOTAL .PARAM NAMES        0      200
TOTAL .PARAM USES         0      400
.MACRO DEFINITIONS        0       50
.MACRO EXPANSIONS         0      200
```

Figure 12-6   (continued)   (c) .PC VERIFY option.

```
** A S P E C **                        DATE 09/23/82    12:30:09
AC ANALYSIS OF CMOS OPERATIONAL AMPLIFIER
TEMPERATURE =  25.0
```

```
  16 ITERATIONS


NODE VOLTAGES -

  NODE    VOLTAGE      NODE   VOLTAGE      NODE   VOLTAGE      NODE    VOLTAGE
 (   1)   5.00000     (   2)  3.34070     (   3)  2.12417     (   4)  -3.89418
 (   5)  -3.89418     (   6)  0.22678     (   9) -5.00000     (  10)   0.0
 (  11)   0.0


V AND I SOURCE CURRENTS -

NAME          CURRENT (MA)       POWER (MW)

VBB           -0.27583            1.3791
VCC            0.27583            1.3791
VIN+           0.0                0.0
VIN-           0.0                0.0


ELEMENT CURRENTS -

NAME          CURRENT (MA)       POWER (MW)

C1             0.00000            0.0000
CL             0.00000            0.0000

ELEMENT POWER DISS. =   0.0000 MW


MOSFET OPERATING POINTS -

NAME      ID (MA)    IG (UA)     VDS       VGS       VSB      POWER (MW)

MP1      -0.06627    0.0       -1.6593    -1.659    0.0         0.1
MP2      -0.07529    0.0       -2.8758    -1.659    0.0         0.2
MP3      -0.03765    0.0       -6.0183    -2.124   -2.8758      0.2
MP4      -0.03765    0.0       -6.0184    -2.124   -2.8758      0.2
MP8      -0.13427    0.0       -4.7732    -1.659    0.0         0.6
MN5       0.06627    0.0        8.3407     8.341    0.0         0.6
MN6       0.03765    0.0        1.1058     1.106    0.0         0.0
MN7       0.03765    0.0        1.1058     1.106    0.0         0.0
MN9       0.13427    0.0        5.2268     1.106    0.0         0.7

MOSFET POWER DISS. =   2.7583 MW

TOTAL POWER DISSIPATION =   2.7583 MW


MOSFET OPERATING POINTS(2)-

NAME          VT        VDSAT      BETA

MP1          1.017      0.467     0.45D-03
MP2          1.019      0.464     0.51D-03
MP3          1.750      0.316     0.64D-03
MP4          1.750      0.316     0.64D-03
MP8          1.005      0.476     0.87D-03
MN5          1.010      6.374     0.29D-05
MN6          0.878      0.154     0.21D-02
MN7          0.878      0.154     0.21D-02
MN9          0.815      0.198     0.47D-02
```

Figure 12-6 (continued) (*d*) dc operating points.

** A S P E C **

AC ANALYSIS OF CMOS OPERATIONAL AMPLIFIER

TEMPERATURE = 25.0

AC MOSFET MODELS -

NAME	GDS	GDD GDG	CGD GDB	CGS CGB	CSB CDB
MP1	-2.895D-04	8.186D-07 2.089D-04	3.078D-14 7.982D-05	7.584D-13 1.597D-13	2.437D-13 6.512D-14
MP2	-3.299D-04	3.192D-07 2.381D-04	9.823D-14 9.150D-05	6.886D-12 1.489D-12	1.748D-12 1.873D-13
MP3	-2.495D-04	5.703D-07 2.015D-04	6.783D-14 4.742D-05	2.288D-12 3.510D-13	3.890D-13 1.001D-13
MP4	-2.495D-04	5.703D-07 2.015D-04	6.783D-14 4.742D-05	2.288D-12 3.510D-13	3.890D-13 1.001D-13
MP8	-5.860D-04	1.628D-06 4.154D-04	5.928D-14 1.689D-04	1.461D-12 3.079D-13	4.688D-13 1.013D-13
MN5	-2.155D-05	3.022D-08 1.898D-05	5.697D-16 2.539D-06	7.520D-14 5.537D-15	4.157D-15 2.113D-16
MN6	-5.246D-04	5.077D-06 3.322D-04	3.433D-14 1.873D-04	5.028D-13 7.578D-14	7.317D-14 1.930D-14
MN7	-5.246D-04	5.077D-06 3.322D-04	3.433D-14 1.873D-04	5.028D-13 7.578D-14	7.317D-14 1.930D-14
MN9	-1.735D-03	1.420D-05 9.295D-04	7.336D-14 7.909D-04	1.075D-12 1.667D-13	1.552D-13 3.042D-14

Figure 12-6 (continued) (*e*) .PC ACVER option.

** A S P E C - M **

AC ANALYSIS OF CMOS OPERATIONAL AMPLIFIER

TEMPERATURE = 25.0

LINEAR AC ANALYSIS -

FREQUENCY	DB VOUT * 1.0D+00 6 0	DB V3 * 1.0D+00 3 0	DB V4 * 1.0D+00 4 0	DB V5 * 1.0D+00 5 0
1.00000D+02	66.35477	-8.79116	-11.42471	30.98085
1.25890+02	66.35328	-8.79088	-11.42442	30.97936
1.58490+02	66.35091	-8.79042	-11.42397	30.97699
1.99530+02	66.34716	-8.78971	-11.42326	30.97324
2.51190+02	66.34122	-8.78858	-11.42213	30.96731
3.16230+02	66.33183	-8.78679	-11.42035	30.95792
3.98110+02	66.31699	-8.78397	-11.41755	30.94307
5.01190+02	66.29356	-8.77956	-11.41315	30.91965
6.30960+02	66.25669	-8.77266	-11.40628	30.88280
7.94330+02	66.19890	-8.76198	-11.39565	30.82501
1.00000+03	66.10884	-8.74568	-11.37941	30.73497
1.25890+03	65.96983	-8.72129	-11.35511	30.59600
1.58490+03	65.75824	-8.68588	-11.31983	30.38446
1.99530+03	65.44268	-8.63665	-11.27079	30.06898
2.51190+03	64.98525	-8.57219	-11.20657	29.61167
3.16230+03	64.34650	-8.49412	-11.12879	28.97312
3.98110+03	63.49442	-8.40797	-11.04295	28.12135
5.01190+03	62.41435	-8.32219	-10.95747	27.04178
6.30960+03	61.11434	-8.24505	-10.88061	25.74257
7.94330+03	59.62215	-8.18181	-10.81758	24.25163
1.00000+04	57.97609	-8.13377	-10.76970	22.60755
1.25890+04	56.21522	-8.09935	-10.73539	20.84983
1.58490+04	54.37317	-8.07571	-10.71183	19.01277
1.99530+04	52.47586	-8.05994	-10.69612	17.12336
2.51190+04	50.54193	-8.04962	-10.68584	15.20190
3.16230+04	48.58414	-8.04296	-10.67919	13.26381
3.98110+04	46.61097	-8.03868	-10.67494	11.32169
5.01190+04	44.62799	-8.03595	-10.67221	9.38744
6.30960+04	42.63874	-8.03419	-10.67047	7.47435
7.94330+04	40.64552	-8.03305	-10.66934	5.59915
1.00000+05	38.64976	-8.03229	-10.66859	3.78412
1.25890+05	36.65239	-8.03173	-10.66805	2.05859
1.58490+05	34.65396	-8.03125	-10.66762	0.45891
1.99530+05	32.65481	-8.03077	-10.66719	-0.97490
2.51190+05	30.65513	-8.03017	-10.66667	-2.20661
3.16230+05	28.65499	-8.02933	-10.66597	-3.21398
3.98110+05	26.65435	-8.02808	-10.66493	-3.99642
5.01190+05	24.65307	-8.02617	-10.66336	-4.57536
6.30960+05	22.65090	-8.02324	-10.66096	-4.98681
7.94330+05	20.64737	-8.01878	-10.65734	-5.27129
1.00000+06	18.64175	-8.01212	-10.65202	-5.46611
1.25890+06	16.63289	-8.00247	-10.64448	-5.60192
1.58490+06	14.61902	-7.98910	-10.63444	-5.70241
1.99530+06	12.59743	-7.97169	-10.62226	-5.78584
2.51190+06	10.56398	-7.95079	-10.60959	-5.86692
3.16230+06	8.51240	-7.92811	-10.59982	-5.95912
3.98110+06	6.43317	-7.90624	-10.59815	-6.07717
5.01190+06	4.31216	-7.88771	-10.61113	-6.24003
6.30960+06	2.12917	-7.87414	-10.64644	-6.47388
7.94330+06	-0.14279	-7.86578	-10.71327	-6.81423
1.00000D+07	-2.53632	-7.86159	-10.82341	-7.30608

Figure 12-6 (continued) (f) .PRINT output.

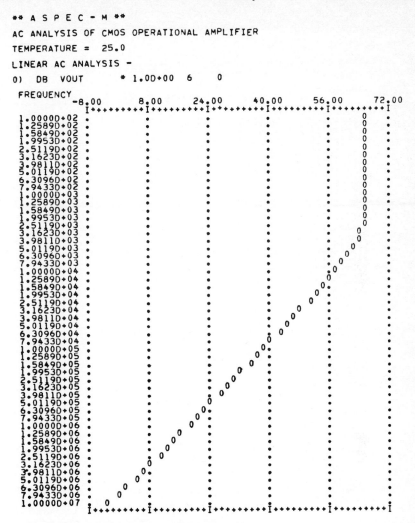

Figure 12-6 (continued) (*g*) .PLOT DB output.

```
** A S P E C - M **
AC ANALYSIS OF CMOS OPERATIONAL AMPLIFIER
TEMPERATURE =  25.0
LINEAR AC ANALYSIS -
```

```
0) PHAS VOUT        * 1.0D+00   6      0

FREQUENCY
         -180.00    -140.00    -100.00     -60.00     -20.00      20.00
      I+++++++++I+++++++++I+++++++++I+++++++++I+++++++++I+++++++++I
1.0000D+02  .          .          .          .          .      O   .
1.2589D+02  .          .          .          .          .      O   .
1.5849D+02  .          .          .          .          .       O  .
1.9953D+02  .          .          .          .          .       O  .
2.5119D+02  .          .          .          .          .       O  .
3.1623D+02  .          .          .          .          .       O  .
3.9811D+02  .          .          .          .          .       O  .
5.0119D+02  .          .          .          .          .      O   .
6.3096D+02  .          .          .          .          .     O    .
7.9433D+02  .          .          .          .          .    O     .
1.0000D+03  .          .          .          .          .   O      .
1.2589D+03  .          .          .          .          .  O       .
1.5849D+03  .          .          .          .          . O        .
1.9953D+03  .          .          .          .          .O         .
2.5119D+03  .          .          .          .        O O.         .
3.1623D+03  .          .          .          .      O    .         .
3.9811D+03  .          .          .          .    O      .         .
5.0119D+03  .          .          .          .  O        .         .
6.3096D+03  .          .          .          .O O        .         .
7.9433D+03  .          .          .        O O.          .         .
1.0000D+04  .          .          .       O   .          .         .
1.2589D+04  .          .          .      O    .          .         .
1.5849D+04  .          .          .     O     .          .         .
1.9953D+04  .          .          .    O      .          .         .
2.5119D+04  .          .          .   O       .          .         .
3.1623D+04  .          .          .   O       .          .         .
3.9811D+04  .          .          .   O       .          .         .
5.0119D+04  .          .          .   O       .          .         .
6.3096D+04  .          .          .  O        .          .         .
7.9433D+04  .          .          .  O        .          .         .
1.0000D+05  .          .          .  O        .          .         .
1.2589D+05  .          .          .  O        .          .         .
1.5849D+05  .          .          .  O        .          .         .
1.9953D+05  .          .          .  O        .          .         .
2.5119D+05  .          .          . O         .          .         .
3.1623D+05  .          .          . O         .          .         .
3.9811D+05  .          .          . O         .          .         .
5.0119D+05  .          .          . O         .          .         .
6.3096D+05  .          .          .O          .          .         .
7.9433D+05  .          .          .O          .          .         .
1.0000D+06  .          .          .O          .          .         .
1.2589D+06  .          .          O           .          .         .
1.5849D+06  .          .         O.           .          .         .
1.9953D+06  .          .        O .           .          .         .
2.5119D+06  .          .     O O  .           .          .         .
3.1623D+06  .          .  O O     .           .          .         .
3.9811D+06  .          .O O       .           .          .         .
5.0119D+06  .        O O          .           .          .         .
6.3096D+06  .     O O             .           .          .         .
7.9433D+06  .   O O               .           .          .         .
1.0000D+07  O                     .           .          .         .
      I+++++++++I+++++++++I+++++++++I+++++++++I+++++++++I+++++++++I
```

```
- END OF RUN -
```

Figure 12-6 (**continued**) (*h*) .PLOT PHAS output.

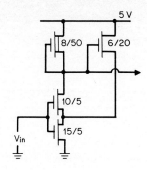

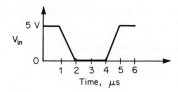

Figure 12-7 Prob. 1.

cutoff frequency (0 dB gain) at 7.9 MHz, and a phase shift of $-132°$. This represents a phase margin of 48°.

A condensed user's manual for the ASPEC program (Version 7.0) is presented in the appendix following this chapter.

PROBLEMS

1. Construct the input file and run a transient analysis of the circuit shown in Fig. 12-7.

Appendix
ASPEC User's Manual, Version 7.0* (Abridged)

A. GENERAL INFORMATION

Program Description

ASPEC is a computer program designed to perform nonlinear dc, nonlinear transfer function, nonlinear transient and linear ac simulation of circuits containing independent sources, linear elements and nonlinear devices.

Types of Analysis

Nonlinear transient analysis predicts the circuit response versus time. Energy storage effects of capacitors and inductors are simulated. The user may print tables or line printer plots of voltages or currents with respect to time. A "snapshot" capability is also available to print out the complete operating state of the circuit at selected time points.

Linear ac analysis predicts the frequency response of the circuit to "impulse" source inputs. The linearized models of nonlinear devices are automatically calculated by the program based upon the dc state of the circuit. Tables and line printer plots of selected complex outputs can be printed versus frequency.

Sources, Elements, and Devices

Independent voltage and current sources may be dc, piecewise-linear, piecewise-exponential and sinusoidal with respect to time for dc and transient analysis. For linear ac analysis, sources may have a finite ac amplitude.

Allowed linear elements are resistors, capacitors.

Nonlinear device types are p-n diodes and MOSFETs. The device model structures and equations are generally fixed; the user defines values to be entered into the device model equations. The more parameters the user specifies, the more complex the model used by the program.

Input Data Organization

Data input to the program consists of a card deck or data file with one component description or execution control statement per card or line. The first card in the data deck is the Title Card which is used to identify the output. The last card must be an .END card. In general, data cards within the data deck may be put in any order. Blank cards will be ignored.

Data may be put anywhere on the card or line, from column 1 to column 80, and need not start in column 1. However, format rules for the *order* in which data is put on each card must be followed.

Instruction words, names and numbers on each card are separated by either one or more blank spaces, a comma (,), an equals sign (=), an asterisk (*) or a slash

*© COPYRIGHT 1982 by Control Data Corporation

(/). If two or more of the non-blank delimiters are used in sequence, a value of zero is implied. For example, the sequence: 1K,,1E-3 defines three numbers, the second of which will be assigned a value of zero. Note that non-blank delimiters must immediately follow the name or number with no intervening blank spaces.

There are no restrictions on the number of blank spaces between words and/ or numbers.

Types of Cards

There are basically two types of cards in each data deck: Component (source, element or device) cards and Instruction Cards.

Component cards start with a unique letter indicating the type of component (e.g. V-source, Resistor, MOSFET, etc.). Example Component cards are V+, R12, MA2B2, Q23-24.

Instruction cards are identified by unique names beginning with a period. The period serves to avoid any name conflicts with possible Component names. Example Instruction cards are .PRINT, .FREQ, .TEMP.

In addition, there are the Title card, the End card, Comment cards and Continuation cards. These cards are described separately.

Names

Component, model and parameter names may be from one to ten alphanumeric characters in length but *must* begin with an alphabetic character (A to Z). Names must not contain embedded blanks, commas, equals signs, asterisks, slashes or dollar signs (, = * / $).

Node Numbers

Node numbers are entered as simple integers without a decimal point. Nodes may be assigned any arbitrary number from 0 to 999, but the ground (datum) node *must* be assigned the number zero.

Note that the node numbers need not be sequential.

Numerical Values

Numerical data values may be entered as integers (no decimal point), numbers with a decimal point (e.g. 1.23, .00123), numbers with an integer exponent (e.g. 25.6E-3, 4E3) or numbers with a concatenated scaling factor (e.g. .15MA, 2PF, 4.3K).

Allowed scaling factors are:

P = pico = 1E-12   V, A, F, H, S, Z, Y, E = 1

N = nano = 1E-9   K = kilo = 1E3

U = micro = 1E-6   MEG = mega = 1E6

M = milli = 1E-3   G = giga = 1E9

All scaling factors following a number are recognized *and cumulative*. Thus 1.5UUF = 1.5PF = 1.5E-12, 4KK = 4E6 = 4MEG, etc. Note that there must be no spaces between the number and the scaling factor letters.

Units of Measurement

Units of measurement are volts, amperes, seconds, ohms, hertz, farads, henries and microns. Unless otherwise indicated by the program, the calculated voltage is output in volts, current is output in milliamperes and power is output in milliwatts.

The. MILS card, if used, defines mils, instead of microns, as the standard unit of length.

Title Card

The first card in each data deck must be the Title Card. All information on this card will be printed verbatim as a page header on the output listing. Thus, the card should contain the user's name and a brief description of the circuit.

Example:

JOHN SMITH   FEEDBACK   AMP   WITH   AV = 10

.End Card

The last card in a data deck must be an Instruction card beginning with the word ".END"

Comment Cards

If a dollar sign ($) is put on a data card, all card characters following the $ will be printed but otherwise ignored by the program. Thus, if the first character on a card is a dollar sign, the entire contents of the card will be printed verbatim but otherwise ignored. Comments can be useful for noting special features of the data.

Examples:

$ COMPENSATION   CAPACITOR   C5   SET   TO   10PF

C5   5   9   10PF $$ CAUSES ROLL-OFF   AT   145   KHZ $$

Continuation Cards

If the first nonblank character on a data card is a plus (+) or an ampersand (&), the card will be assumed to be a continuation of the previous data card. Individual names and numbers must not be split between two cards.

There is no limit on the number of continuation cards that may be used in sequence.

B.   RESISTOR, CAPACITOR AND INDUCTOR

Linear resistor, capacitor and inductor elements are specified using the following formats:

Rxxxxxxxxx n1 n2 value

Cxxxxxxxxx n1 n2 value

Lxxxxxxxxx n1 n2 value

a. "Rxxxxxxxxx" indicates a resistor element.
"Cxxxxxxxxx" indicates a capacitor element.
"Lxxxxxxxxx" indicates an inductor element.
(The "xxxxxxxxx" represents an optional name of from one to nine alphanumeric characters). To prevent ambiguity in the data, each element name should be unique.

b. "n1   n2" represent the node numbers of the element. The defined reference direction for positive branch current is from node n1 to node n2.

c. "value" represents the nominal resistance, capacitance or inductance of the element at a temperature of 25 degrees C. Zero-valued resistors and inductors are *not* allowed, but zero-valued capacitors may be specified.

Examples:

R   2   3   1K

REXT   32   42   200K

C23AB   1   201   .15PF

RX   2   3   1K

CX   5   4   2PF   .0001

C.   VOLTAGE AND CURRENT SOURCES

For dc and transient analysis, sources can be dc or piecewise-linear. For linear ac analysis, an ac amplitude may be specified.

DC Source

A dc source remains constant for all time values. The general formats for dc sources are:

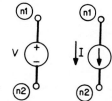

Vxxxxxxxxx n1 n2 DC vdc

Ixxxxxxxxx n1 n2 DC idc

a. "Vxxxxxxxxx" indicates a voltage source.

"Ixxxxxxxxx" indicates a current source.

(The "xxxxxxxxx" represents an optional name of from one to nine alphanumeric characters).

b. "n1   n2" represent the node numbers of the source. For voltage sources, the voltage at node n1 equals the voltage at node n2 plus the source value. Positive V-source and I-source current is defined as going from node n1 to node n2 through the source.

c. "DC" indicates that the dc source value follows. For a simple dc source, the letters "DC" are optional and may be omitted (see examples below).

d. "vdc" and "idc" represent the dc value of the voltage or current source in volts or amps, respectively. Source values may be positive, negative or zero. If not specified, vdc = 0 and idc = 0 are assumed.

Examples:

VDD   5   0   15V

IS   14   15   DC   1MA

V23   3   23   DC   -5V

IEQ2   14   28   1.5UA

Piecewise-Linear Source

A piecewise-linear source is defined with value-time "breakpoint pairs." Between each of the breakpoint pairs, the source value is linearly interpolated. The general formats for piecewise-linear sources are:

Vxxxxxxxxx   n1   n2   DC   vdc   PL   v1   t1   v2   t2   v3   t3
+   v4   t4   . . .   vn   tn

Ixxxxxxxxx   n1   n2   DC   idc   PL   i1   t1   i2   t2   i3   t3
+   i4   t4   . . .   in   tn

a. "Vxxxxxxxxx" and "Ixxxxxxxxx" indicate a voltage or current source. ("xxxxxxxxx" represents an optional name).

b. "n1   n2" are the source nodes. The defined reference directions for current are the same as for dc sources.

c. "DC   vdc" or "DC   idc" defines the value of the source at time = 0. If the value is not specified, vdc = 0 or idc = 0 is assumed.

d. "PL" indicates that value-time breakpoint pairs follow.

e. "v1   t1   v2   t2   v3   t3   . . .   vn   tn" represent voltage-time breakpoint pairs. If t1 = 0, the v1 value will override any specified vdc value.

"i1 t1 i2 t2 i3 t3   . . . in tn" represent current-time breakpoint pairs. If t1 = 0, the i1 value will override any specified idc value.

Note that t1 < t2 < t3 < . . . < tn must hold true for all time values. After time tn, a constant value for vn or in is assumed.

Breakpoint pairs may be continued onto following cards, but a given pair *must not* be split between two cards.

Refer to the examples below.

Examples:

```
VIN  12  0  DC  0V  PL  0V  2NS  5V  2.5NS  5V  4NS  0V  4.5NS
+    0V  6NS  5V  6.5NS
```

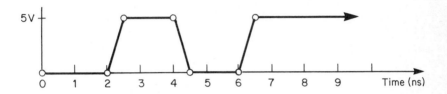

(Note that the value remains constant after tn = 6.5 ns)

AC Source Value

For linear ac analysis, the ac source values default to zero unless otherwise specified (i.e. voltage source nodes are grounded and current sources are effectively removed). If a source is to have a finite ac value, it must be specified on the source card following the letters "AC".

The general formats are as follows:

Vxxxxxxxxx n1 n2 AC vac

Ixxxxxxxxx n1 n2 AC iac

a. "Vxxxxxxxxx" and "Ixxxxxxxxx" indicate a voltage or current source. (The "xxxxxxxxx" represents an optional name.)

b. "n1 n2" are the source nodes.

c. "AC vac" specifies the ac voltage impulse value
"AC iac" specifies the ac current impulse value.
Normally, the values vac = 1 and iac = 1 should be used in order to simplify interpretation of the ac output. If not specified, vac = 0 or iac = 0 is assumed.

The ac values may be specified in conjunction with dc, piecewise linear, exponential or sinusoidal values (refer to the examples below).

Examples:

VAC   2   0   DC   2V   AC   1

I22   4   5   AC   1   DC   1.5MA

VQ2   6   12   DC   2V   PL   2V   20NS   0V   40NS   2V   60NS   AC   1

IAC2   11   12   DC   .5MA   AC   .5MA

D. DEVICE MODELS

MOS Field-Effect Transistors

The general format for MOSFET device specification is:

Mxxxxxxxxx nd ng ns mname W L

or

Mxxxxxxxxx nd ng ns nb mname W L

a. "Mxxxxxxxxx" indicates a MOS Field Effect Transistor. (The "xxxxxxxxx" represents an optional name of from one to nine alphanumeric characters.)

b. "nd ng ns nb" are the drain, gate, source and bulk nodes, respectively, of the MOSFET. If the bulk (substrate) node is not specified on the device card, then its value is obtained from the .MODEL card.

c. "mname" represents the 1 to 10 character model name. This name is used to refer to a specific .MODEL card. (ref. *DEVICE MODELS*).

d. "W L" represent the values of the *drawn* width and length, respectively, of the device. These values must be specified.

Examples:

M   1   2   3   MOS   1.   .2

M15   22   45   6   MODL   1.5   .3

MABC3   4   9   98   NMODL   8   .25

MGSW   5   17   19   55   NYC   23   3.5

The .MODEL Card

The .MODEL card is used to define the general properties of each different type of device. Diode, BJT, JFET and MOSFET devices all have different model parameter names and values. . . . However, the general format is the same for all. It is:

.MODEL   mname   type   pname = pval   pname = pval   pname = pval
+   pname = pval   pname = pval   . . .

a. ".MODEL" indicates a model specification card.

b. "mname" represents the 1 to 10 character user-defined name assigned to the model. This is the name referenced by the individual device cards.

c. "type" represents one of ten possible device type names. Three are defined as follows:

D   PN diode

NMOS   N-channel MOSFET

PMOS   P-channel MOSFET

d. "pname = pval   pname = pval . . ." represents a list of model parameter names followed by their values. The parameters need not be specified in any particular order. Allowed parameter names for one type of device is given below.

Examples:

.MODEL   DIODE   D   IS = IE-15   N = 1.04   RS = 10.

.MODEL   MOS   NMOS   BULK = 1   VT = 2.5   UB = 650   BETA = 15U
+   COX = 3.45E-16   LATD = .5

MOSFET Parameter Names*

Name	Parameter type	Units	Default
VT	Threshold voltage with zero backbias	V	10^6
UB	Low field (bulk) mobility	$cm^2/(V \cdot s)$	750 N-ch
			250 P-ch
COX	Gate oxide capacitance	$F/\mu m^2$	†
TOX	Gate oxide thickness	Å	690
DNB	Bulk doping concentration	cm^{-3}	10^{15}
METO	Metal overlap	μm	0.
ESAT	Velocity saturation critical field	V/cm	infinity
VMX	Scattering limited velocity	cm/s	infinity
LATD	Lateral out-diffusion on each side	μm	0.
LDEL	Channel length increment	μm	0.
LMLT	Channel length scaling factor		1.
WDEL	Channel width increment	μm	0.
WMLT	Channel width scaling factor		1.
BETA	Transconductance	A/V^2	†

*Note: Device models are continually being updated. Consult with Information System Design for latest complete device model.

†Calculated internally.

E. ANALYSIS SPECIFICATION

Types of Analysis Cards

The .TRAN and .FREQ cards are used in running ASPEC to request nonlinear transient . . . and linear ac analysis, respectively. Note that these cards only request that the program perform a specific type of analysis; they *do not* specify how the outputs are to be printed. Outputs are requested using the .OP, .PRINT, .PLOT and .OUTPUT cards described in the next section. Thus, a given type of analysis will only be performed if both the analysis card (.TRAN or .FREQ) and the corresponding output card (.PRINT, .PLOT or .OUTPUT) are specified.

One each nonlinear transient and linear ac analysis may be performed on the same run. If more than one each .TRAN, or .FREQ card is used in a given data deck, the program will use the last card of each type.

The program always performs a dc analysis prior to nonlinear transient or linear ac analysis.

.TRAN Card

The .TRAN card is used to control the printing timestep during transient analysis. It must be used in conjunction with .OP, .PRINT, .PLOT or .OUTPUT cards in order to generate any transient printout.

The general format for this card is:

.TRAN   d1   t1   d2   t2   d3   t3   . . .   dn   tn

a. ".TRAN" requests that a transient analysis be performed. DC analysis will be automatically performed to determine the starting conditions. Only one .TRAN card is allowed per run.

b. "d1   t1   d2   t2   d3   t3   . . ." represent timestep and timepoint value pairs.

During transient analysis, tabular and/or plotted transient outputs will be printed from time = 0 to time = t1 in increments of d1. Outputs will then be printed from time = t1 to time = t2 in increments of d2. Outputs are then printed in increments of d3 up to t3. Printing continues in this fashion up until the last timepoint.

Continuation cards may be used, but a timestep/timepoint pair must not be split between two cards.

Examples:

.TRAN   1NS   100NS

.TRAN   1NS/10NS   10NS/100NS   1NS/110NS   10NS/200NS
+   1NS/210NS   10NS/300NS

.TRAN   100NS   1US   1NS   1.02US

Note: The .TRAN card only controls the timestep used for printout. During transient analysis, the program may use considerably smaller timesteps than those specified for printout.

.FREQ Card

The .FREQ card is used to control the frequency print points during linear ac analysis. It must be used in conjunction with .PRINT, .PLOT or .OUTPUT cards in order to generate any frequency response printouts.

The general format for this card is:

.FREQ typ1 i1 fa1 fb1 typ2 i2 fa2 fb2 ... typn in fan fbn

a. ".FREQ" requests that an ac analysis be performed. DC analysis will be automatically performed to determine the linear ac models. Only one .FREQ card is allowed per run.

b. "typ1", "typ2", etc. represent any of the optional words "DEC", "OCT" or "LIN" which define the type of frequency increment.

DEC: Causes i points per *decade* to be printed logarithmically between frequencies fa and fb

OCT: Causes i points per *octave* to be printed logarithmically between frequencies fa and fb

LIN: Causes a *total* of i points to be printed linearly between frequencies fa and fb

If "typ1" is not specified, "DEC" is assumed. If any "typ2" through "typn" is not specified before a given "i fa fb" set of values, the program will assign the same type as used for the preceding output (if any) on the card.

c. "i1 fa1 fb1", "i2 fa2 fb2", etc. represent the values for the number of frequency points per DEC, OCT or LIN, and the start and finish frequency values as described above. The frequency values are arbitrary as long as each fa < fb. Also, a given fa must be greater than zero if the increment type is "DEC" or "OCT".

Continuation cards may be used, but a "typ i fa fb" group must not be split between two cards.

Examples:

.FREQ 10 1K 1MEG

.FREQ LIN 100 1K 1MEG 200 1MEG 1.1MEG
+200 10K 11K 50 11K 12K

.FREQ DEC 10 1 1K OCT 10 10 1K DEC 5 1K 100K

F. OUTPUT SPECIFICATION

The .OP, .PRINT, .PLOT and .OUTPUT cards are used to define which specific node voltages, branch currents or power dissipations are to be printed and/or plotted for dc, nonlinear transient or linear ac analysis.

Although there is a limit on the number of different outputs, multiple uses of an identical output count only as a single output. Thus, if the same output is to be both printed and plotted, or plotted more than once, it is still considered to be only one of the total allowable outputs.

The .OP card can be used to print the complete operating condition of the circuit in the dc state and at selected points during transfer function or transient analysis.

For each .PRINT card, a tabular listing of the specified outputs will be printed versus time, a parameter value, temperature or frequency.

For each .PLOT card, a line printer plot containing the waveforms of the specified outputs will be printed versus time, a parameter value, temperature, or frequency.

For each .OUTPUT card, the numerical value of one output will be printed alongside a plot of the specified outputs.

The formats for the output specification cards are given on the following pages.

.OP TRAN Card

The .OP TRAN card is used to print out the complete operating condition of the circuit in the quiescent (dc) state and at selected points during transient analysis.

The general format for this card is:

.OP   TRAN   typ1   t1   typ2   t2   typ3   t3   ...   typn   tn

a. ".OP   TRAN" indicates a dc and transient operating point print card. The word "TRAN" can be omitted (see examples).

b. "typ1", "typ2", "typ3," etc. represent any of the optional words "ALL", "VOL" or "CUR" which define the amount of printout at each timepoint as follows:

ALL: Print node voltages and component currents and power

VOL: Print node voltages only

CUR: Print currents and power dissipations only

If a type is not specified, "ALL" is assumed.

c. "t1", "t2", "t3", etc. represent the timepoints at which the operating conditions are to be printed, where t1 < t2 < t3 < ... < tn. If the dc operating conditions are to be printed, then t1 = 0 should be specified. If no time values are specified, t1 = 0 is assumed.

Note that transient points will not be printed beyond the last timepoint specified on the .TRAN card.

Continuation cards may be used but a "typn tn" pair must not be split between two cards.

Examples:

.OP

.OP 2US 3US 6US

.OP TRAN VOL 0NS VOL 10NS VOL 30NS VOL 40NS

.OP ALL 0NS CUR 10NS VOL 100NS ALL 200NS
+ CUR 300NS ALL 400NS

Transient Analysis Outputs

For transient analysis, the general forms for output specification are:

.PRINT Vxxxxxxx n1 n2 Ixxxxxxx name Pxxxxxxx name

.PLOT Vxxxxxxx n1 n2 Ixxxxxxx name Pxxxxxxx name

a. ".PRINT" indicates that the outputs on the card are to be printed in tabular form.

".PLOT" indicates that the outputs on the card are to be printed as line printer plots. The plot coordinates will be automatically calculated by the program.

b. "Vxxxxxxx n1 n2" specifies that the voltage between two nodes, numbered n1 and n2, is to be printed or plotted. (The "xxxxxxx" represents an optional output name of up to nine characters.) Voltage output may be requested between any two nodes in the circuit.

c. "Ixxxxxxx name" specifies that the current through source, element or device "name" is to printed or plotted. ("xxxxxxx" represents an optional output name.) If more than one circuit component has the same name (this should be avoided), the output component used will be the one that appears first in the data deck.

For current output through a MOSFET, the first two characters of the output name should be ID, IG, IS or IB, requesting drain, gate, source or bulk current, respectively. If the second character is not a D, G, S or B for JFET or MOSFET output, the drain current will be printed.

Reference directions for *positive* current output are as follows:

V and I sources: From the first node to the second node (n1 to n2).

R, C: From the first connecting node to the second (n1 to n2).

Diodes: From the p-type node to the n-type node (np to nn).

MOSFETs: Drain, Gate and Bulk current entering the device; Source current leaving.

d. "Pxxxxxxx name" specifies that the power dissipation in the component "name" is to be printed or plotted. ("xxxxxxx" represents an optional output name.) Power output in Diodes, and MOSFETs includes the dissipation in the capacitors and extrinsic resistors. All characters following the "P" are ignored.

V, I and P outputs may be put in any order or combination on the .PRINT, .PLOT and .OUTPUT cards. If necessary, outputs may be continued onto following cards, but each individual output specification must not be split between two cards.

Examples:

.OUTPUT   IC1   Q1

.PRINT   V1   1   0   V3   3   0   IC   Q3   I   R5   VOUT   23   24

.PLOT   VOUT   5   0   V21-23   21   23   V46-50   46   50

.PRINT   P   M15   ID   M15   IG   M15   P64   M64   IG   M64
+   P   M40   ID   M40   IG   M40   IB   M40

.OUTPUT   V23   23   0   IEQ53   Q53

Linear AC Analysis Outputs

AC analysis outputs are requested using the .PRINT, .PLOT cards in a manner similar to transient analysis output requests. However, there are two important differences:

1. Only AC voltage and current outputs may be requested. AC power output requests will be ignored.

2. The user must additionally specify which component of the complex ac output (magnitude, phase, db gain, real part or imaginary part) is to be printed.

The output forms for linear AC response .PRINT, .PLOT cards are:

.PRINT   type   Vxxxxxxx   n1   n2   type   Ixxxxxxx   name   . . .

.PLOT   type   Vxxxxxxx   n1   n2   type   Ixxxxxxx   name   . . .

a. ".PRINT" requests tabular output.
   ".PLOT" requests line printer plotted output.

b. "type" represents one of the words, "MAG", "PHAS", "DB", "REAL" or "IMAG", which defines the component of the AC output that is to be printed, as follows:

MAG: Absolute magnitude of the output

PHAS: Phase of the output defined from 0 to 180 degrees and from 0 to
          −180 degrees

DB: 20*log(MAG) of the output

REAL: Real part of the output

IMAG: Imaginary part of the output

The "MAG", "PHAS", "DB", "REAL" and "IMAG" names can be put before each Vxxxxxxx or Ixxxxxxx output to request a different interpretation of each output. If the "type" is not specified before a given V or I output, the program will assign the same type as given to the preceding output (if any) on the card. For example:

.PLOT DB V 1 0 V 2 0 MAG V 3 0 V 4 0 V 5 0

will cause the program to plot the db output voltages for nodes 1 and 2 and the magnitude of the voltages at nodes 3, 4 and 5.

If no "type" is specified for any of the outputs on a .PRINT, .PLOT or .OUTPUT card, the outputs will be used for transient printout only.

c. "Vxxxxxxx   n1   n2" specifies that the ac voltage between nodes n1 and n2 is to be printed or plotted.

d. "Ixxxxxxx   name" specifies that the ac current through component "name" is to be printed or plotted.

"IGxxxxxx", IDxxxxxx", "ISxxxxxx" and "IBxxxxxx" request ac Gate, Drain, Source and Bulk currents in MOSFET devices.

Reference directions for zero-phase ac current output are the same as for positive transient current output.

Examples:

.PRINT REAL V 1 O IMAG V 1 O MAG V 1 O

.OUTPUT DB V22-23 22 23 V26-30 26 30

.PRINT MAG V12 12 0 DB V12 12 0 PHAS V12 12 0
+   MAG V23 23 0 DB V23 23 0 PHAS V23 23 0
+   MAG V25 25 0 V30 30 0 VOUT 48 1

.OUTPUT MAG ID13 J13

G. PRINT CONTROL SPECIFICATION

.PC Card

The .PC card is used to control the amount and format of the program data printout. The general format for this card is:

.PC option option option option

a. ".PC" indicates that print control options follow. Any time this card is used, any previously set options are removed.

b. "option" represents a key word as follows:

BRIEF: Suppresses reprinting of the data card images. All data cards follow-ing the .PC card are affected. Thus, if no data cards are to be printed, the .PC card must be the first card in the data deck after the Title Card.

REPRINT: Requests that the instruction card, source, element, device, and .MODEL card data as interpreted by the program be printed. This reprint will also include all components created by macro expansions. The data reprint allows checking to insure that the input data values were interpreted correctly by the program.

VERIFY: Causes the program to print a node connection table, iteration con-trol parameters and program storage usage. The node connection table shows which circuit components are connected to each node in the circuit.

ACVER: Causes the program to print the linearized Diode, and MOSFET models to be used during the linear ac analysis. ACVER is ignored if no ac analysis is requested.

Examples:

.PC   VERIFY   REPRINT   ACVER

.PC   BRIEF

H.   ANALYSIS CONTROL

.CONTROL Card

The .CONTROL card allows the user to modify the program's dc and transient analysis iteration scheme. It should be used with considerable discretion since changes in the standard values could result in greatly increased execution times with no appreciable improvement in accuracy.

The general format for this card is:

.CONTROL   option = value   option = value   option = value   . . .

a. ".CONTROL" indicates the type of card.

b. "option" represents the name of an iteration control option whose default value is to be changed.

c. "value" represents the numerical value to be assigned to the given control option.

Given below are the option names, their default values and a brief description of each.

ABSV = 50UV   Absolute voltage change tolerance

RELV = .001   Relative voltage change tolerance

ABSI = 1NA   Absolute current change tolerance

RELI = .05   Relative current change tolerance

RMAX = 1   Maximum time to print step ratio

RMIN = 1E-3   Minimum time to print step ratio

IMAX = 8   Maximum iteration limit

IMIN = 3   Minimum iteration count

I. ALTER AND RERUN SPECIFICATION

.ALTER Card

The .ALTER card allows the simulation to be rerun several times with selected changes on each run.

The card has the form:

.ALTER

The .ALTER card is placed at the end of the original data deck (in place of the .END card). The .ALTER card is then followed by source, element, device or instruction cards. These cards can be followed by either an .END card or another .ALTER card.

For the first run, the program will look only at the data up to the first .ALTER card. After the first simulation has been performed, the program looks at the cards between the .ALTER card and the next .ALTER or .END card.

There is no limit to the number of times an .ALTER card may be used in a data file. However, each time an .ALTER card is used, the entire simulation will be rerun, which could result in excessively long execution times. The .ALTER feature is intended primarily for convenience when only a few data or instruction cards need to be changed in a large data file.

Examples: In the following example, the first .ALTER card is used to change both the node numbers and the values of resistors R1 and RX3. The second .ALTER card is used to change the voltage of the source V+. Note that on the second rerun, the values and node numbers of resistors R1 and RX3 will be the same as they were on the first rerun.

J. DOE   ALTER   TEST

.OP

V+   2   0   1V

RX3   2   1   1K

R1   1   0   2K

.ALTER

R1   2   0   2K

RX3   2   0   2K

.ALTER

V+   2   0   2V

.END

J.   .MILS Card

The .MILS card, if used, defines mils, instead of microns as the standard unit of length. Placed anywhere in the data deck, a number of selected quantities whose units involve length or area will be affected.

The only quantities not affected are those whose "natural" unit of length is centimeters: doping levels (cm^{-3}) and field strengths (v/cm).

13

Layout, Mask, and Assembly

13-1 LAYOUT AND DESIGN RULES

Layout is the drawing of the actual interconnected circuit as it appears on the integrated circuit. It should show the active and passive devices with the proper layer and structures, and the means to interconnect them. Of course it is imperative that the interconnections be free of errors.

A demonstration[1] of the basic concept of layout is given in Fig. 13-1 for a silicon gate NMOS circuit. Starting at the bottom of the schematic diagram, we call the source of transistor $Q3$ area 1 and note that it is connected by means of a contact hole to the metal ground lead. Lead C forms the input gate of $Q3$. The drain of $Q3$ connects directly to the source of $Q2$ by means of a single diffusion (area No. 2) rather than two diffusion areas connected by a metal strap. A similar case exists for area 3. Area 4 is the output; hence a contact and metal lead are needed to connect to other circuits. Area 4 is also the source for load device QL. The gate of QL connects to that of its source by means of a buried contact. Notice that for QL, the channel length L (from source to drain) is longer than the channel width W. Lastly, the source of device $Q4$ is common with area No. 1. Notice too that in order for the series combination of $Q1$, $Q2$, and $Q3$ to have the same on-resistance as $Q4$, their W/L ratio has to be three times as large.

For a silicon-gate process, the gate area is determined by the intersection of the active area layer with the polysilicon gate layer. Diffusion areas comprise the rest of the active area layer. In other words, polysilicon lines cannot cross diffusion runs without cutting off the diffusion runs. For that reason, input B crosses area 1 with metal. Finally, metal can make contact with both polysilicon and diffusion, whereas a buried contact only allows connection between polysilicon and underlying diffusion.

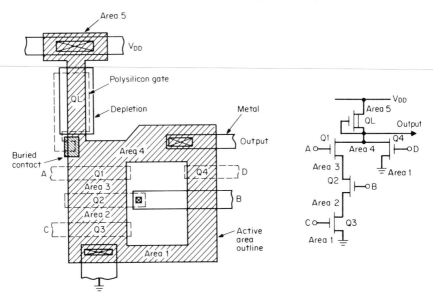

Figure 13-1 Combination NAND-NOR gate with actual layout. *(Concept after Ref.1.)*

Another example of a layout is shown in Fig. 13-2, this time for a CMOS inverter and a two-input NAND gate.[2] The regularity of this results in a very compact layout while still providing easy interconnection with other circuits. One can identify five n-channel devices in the bottom row of the figure and five corresponding p-channel devices in the row above it. The polysilicon gates of the group of three n-channel devices on the left run up vertically to become the gates of three p-channel devices. For the right group of two pairs of n- and p-channel transistors, the vertical gate connection is offset, leaving a single isolated gate p and n channel. The V_{DD} and V_{SS} power supply metal lines run horizontally across (but are insulated from) the top and bottom rows of transistors. Adjacent to the left sides of the two groups of active devices are two vertical polysilicon lines used to route signals under the power supply lines. In between the two groups are small pads of diffusions to contact the substrate with V_{DD} and the p-tub with V_{SS}.

The solid black lines show the interconnections needed to form the logic gates. For the inverter on the left, the two drains are connected and their sources are tied to the proper supply rails. For the NAND gate on the right, one makes use of the fact that sources and drains are interchangeable to allow parallel connections for p channels and series connections for n channels.

Layouts are governed by *design rules* (layout rules) that dictate feature size and dimensional relationships between layers. Such rules reflect the current capability of each process technology and process area. The rules can be quite extensive, involving a document of 10 to 50 pages of dimensioned drawings. The document lists the minimum feature size on each layer and minimum distances between shapes on different layers. However, since chip designers strive for minimum chip size, those minimum dimensions are invariably what is used everywhere.

Design rules tend to fall into the following categories:

1. **Minimum Width** This defines the minimum line widths that can be routinely and reliably patterned without notching or necking of the line. Layers with a thinner film thickness have less trouble being patterned into finer lines.

2. **Minimum Space** This defines dimensions that can be routinely and reliably etched apart with no bridging. On layers such as polysilicon and metal, it is dictated by ability to separate physically two lines of that material. On layers such as diffusion (active area), it is dictated by electrical punch-through.

3. **Minimum Pitch** This is a seemingly superfluous rule that merely sums the minimum width and minimum space and defines that as pitch. However, minimum pitch is the true measure of the photolithographic

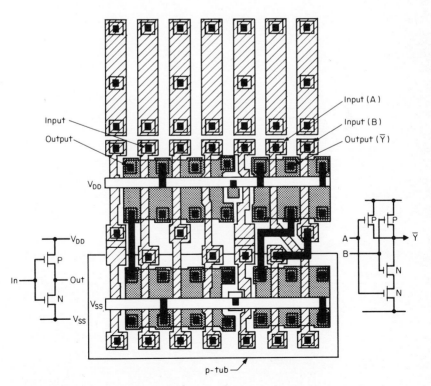

Figure 13-2 Layout of CMOS inverter and two-input NAND gate. *(After Ref. 2.)*

and etching capability of a process. This comes from the fact that minimum space can be played off against minimum width, or vice versa. For example, a very narrow line can be obtained by extreme overetching, but it would not be possible to place two such lines very close.

4. ***Overlap***   This is defined between two layers. For example, a contact to diffusion requires allowances for such things as misalignment between the two layers, enlarging of contact size during etching, and shrinking of diffusion area during field oxide growth. If the diffusion does not overlap the contact by a sufficient amount, metal would short to substrate.

5. ***Separation***   Unrelated features on certain layers must be separated by some minimum amount. Again, this must take into account the worst-case misalignment between layers and shrinkage or enlargement of features during etching.

Misalignment between masking layers plays a key role in determining design rules. Misalignment between two layers that are being directly aligned to each other depends on obvious factors such as operator skill and machine accuracy. However, the way the alignment sequence is set up for a process also affects the resulting accuracy. Figure 13-3 shows an align-

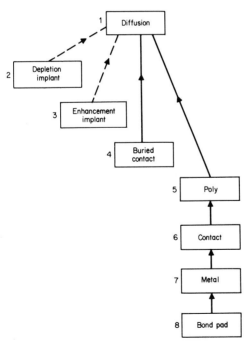

Figure 13-3   Alignment sequence for HMOS process. Dashed line means "leaves no imprint."

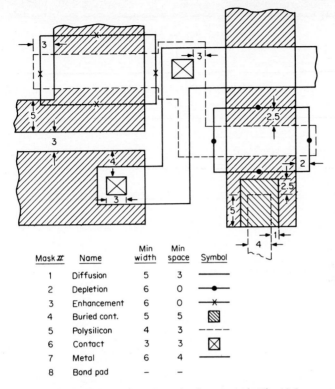

Mask $\mathbb{I}$	Name	Min width	Min space	Symbol
1	Diffusion	5	3	————
2	Depletion	6	0	—•—
3	Enhancement	6	0	—✕—
4	Buried cont.	5	5	◨
5	Polysilicon	4	3	– – – –
6	Contact	3	3	⊠
7	Metal	6	4	————
8	Bond pad	–	–	

Figure 13-4 Example of design rules for process in Fig. 13-3.

ment sequence for a HMOS process. The number beside each layer corresponds to the masking sequence. Layers that are directly aligned to each other, such as poly to diffusion, have the lowest misalignment errors. Layers that are indirectly aligned, such as between buried contact and poly, should allow for larger misalignment.

Uneven wafer expansion and mask "run-out" (unequal step and repeat distance) are other contributing factors to misalignment, but they are usually not dominant.

An example of a very abbreviated set of design rules is given in Fig. 13-4 for a HMOS process.

13-2 FROM LOGIC DESIGN TO MASKS

The sequence of events that takes place from the system-level logic description all the way to a completed mask set ready for wafer processing will be described in this section. The discussion is keyed to various activity blocks presented in flowchart form in Fig. 13-5.

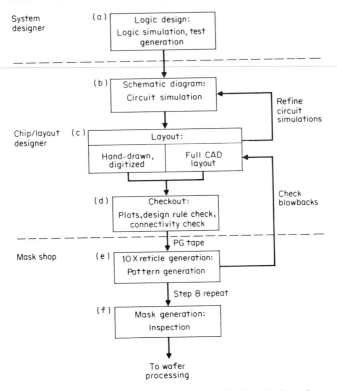

Figure 13-5 **Flowchart from logic design to final mask plates. Letters are keyed to discussion in text.**

(a) Logic Design

A system designer produces a logic schematic that is then turned over to the MOS chip designer. The system designer ensures that the system is free of logic hazards and points out signal propagation paths where race conditions exist. Test sequences should also be generated that can detect a false output whenever there is an internal fault condition where nodes are stuck at a 0 or 1.

(b) Schematic Diagram

The MOS chip designer translates the logic diagram into a transistor-level schematic diagram. This includes specifying the device size (W/L ratio) and circuit interconnect. The techniques for doing this have been covered in Chap. 9. As explained in that chapter, the specification of device size is a result of dc and transient analyses. The transient analysis can be performed on simple circuits by means of graphs or simple equations; but for circuits of typical complexity, the analysis can be done less tediously and with more accuracy with circuit simulation programs on computers (see Chap. 12). The circuit simulation program is put to good use in chasing down race conditions among circuit paths. This necessitates including parasitic capacitances and resistances in the simulation.

(c) Layout

Once the schematic is complete, layout begins. It is a good practice to prepare a floor plan of the whole chip. It should show the general flow of signals from one end of the chip to the other. Each circuit block within the chip should have its area meticulously estimated and its outline fairly well defined. Such a chip plan is indispensable if more than one mask designer (layout designer) is working on the project. Each has to adhere strictly to the boundaries of the block assigned to him or her in order to have proper interface and not waste silicon area.

One aspect of the preplanning is to route the V_{DD} and ground paths in metal except for the very last runs within cells.[3] The metal runs must be quite wide to decrease IR drops and to minimize electromigration problems. One such plan is shown in Fig. 13-6. Notice that G_{ND} and V_{DD} are both available at all sections of the chip and yet do not cross each other.

Layout can take one of two general paths, depending on whether it is manual or fully mechanized. The first method is to draw the layout man-

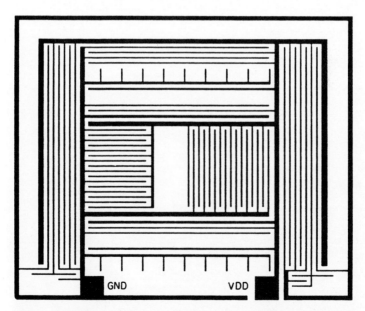

Figure 13-6 V_{DD} and G_{ND} paths preplanned for a chip. (*After Ref. 3. Reprinted with permission.*)

ually on Mylar films. Mylar (with grids imprinted on it) is preferred over paper because of its dimensional stability and ability to withstand repeated erasures. Different layers are drawn with different colored pencils to form a *composite* drawing. Often individual sections of circuits are drawn on "paper dollies" to move around in an attempt to come up with a routing scheme with the smallest area.

At this point, layout information can be fed back to the circuit designer to refine the simulations with more specific information on parasitic effects. This is shown as the first iteration loop in Fig. 13-5.

When layout is complete, it is *digitized* with a large drafting machine that can sense the exact position of a cursor. The cursor is traced over all corners of the shape to be digitized. As the cursor traces over the corners, the coordinates are recorded by simply pressing a button. The various layers on the composite are kept separate by the machine. A very powerful feature that is available is to digitize only a small section and ask the software to take that section and either translate, rotate, reflect about an axis, or repeat to form an array.

The second layout method is fully mechanized and bypasses the intermediate step of digitizing. By using a light pen on a CRT (cathode-ray tube) and with a small digitizing tablet, shapes can be drawn and stored very rapidly (Fig. 13-7). Shapes can be added or deleted and lines moved and rerouted with considerable ease. For a person well-versed in this method of layout, the overall productivity is much higher than the manual method.

Figure 13-7 Fully mechanized layout on CRT screen.

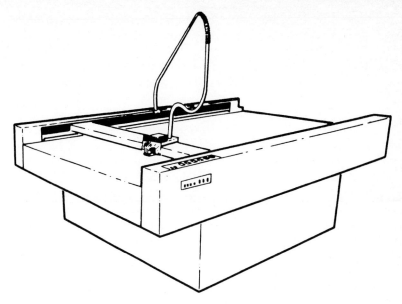

Figure 13-8 Machine for line-plotting IC layout. *(After Ref. 4.)*

It should be pointed out that the digitizing machine is usually part of the fully mechanized system. It is thus possible to mix the two methods. For example, layouts can be manually drawn and digitized and then corrections implemented on the CRT screen.

(d) Checkout

With either the manual or computer-aided system, once the encoding of data is complete, a plot is needed. It is generated on a huge but very precise, stepping-motor driven, *XY* plotter[4] with over 20 ft^2 of plotting area (Fig. 13-8). Either the individual layers or a multicolored composite can be plotted. The magnification is set at 200 to 1000X to allow a close check of critical dimensions. Another means of plotting is electrostatic printing, where an image is formed as paper passes over a linear printhead. The advantage of this method is increased speed and the ability to shade layers. However, the plots are only in black and white.

The process of checking the plots is extremely tedious at best. Not only must the device sizes and the circuit interconnections be without error, the layout has to satisfy all design rules. Consider that typical chips have 5000 to 50,000 transistors and must satisfy 50 to 100 individual design rules. It is small wonder that most companies look to computer-aided checkouts to ensure an error-free layout. Computer programs are available to check for design rule violations. This requires coding all design rules in a format usable by the program. The output consists only of the violations upon which the designer can concentrate his or her attention. More sophisticated programs will compare the digitized data base with a separately entered circuit schematic and flag any differences in connectivity. The schematic can further be compared with the data base used for logic simulation and provide verification from logic to layout. Any one of the programs mentioned above is an extremely heavy user of computing time. However, if a fatal error is prevented, the saving in time and resources outweighs such costs.

After a complete checkout, there is one other activity that is sometimes needed before the data are shipped to the mask-making facility. For a variety of reasons, e.g., when a circuit digitized for a negative photoresist process is switched to positive, certain mask layers of the layout have to be "sized." This refers to enlarging or shrinking by software all features or shapes by a fixed amount on all sides. The relative positions of all the features stay the same. This is contrasted with shrinking, where the overall chip size is reduced.

Final information is transmitted to the mask-making facility in the form of magnetic tapes. This is sometimes referred to as "PG tape" because the layout information has been translated into a format directly readable by the mask-making *pattern generator*.

(e) Reticle Generation

The PG tape drives a pattern generator to produce a 10X reticle. The pattern generator "writes" on the reticle with a beam of light that flashes on and off at a very high rate. The shape of the light beam can be changed on demand during writing to speed up the process.

Once generated, the 10X reticle is magnified by another factor of 15 to 20 to generate 150 to 200X transparency films called "blowbacks." The blowbacks are shipped back to designers for one final check. They are compared with plots of the same scale to ensure that no features are added, deleted, or altered.

(f) Mask Generation

When approved, the reticle is mounted on a step and repeat camera which optically reduces it by a factor of 10 to 1X size and exposes it on the photosensitive mask plate. After each exposure, the plate is translated by a moveable stage a certain X, Y stepping distance. For three to five specific locations on the mask, the circuit reticle is blanked and a reticle for a test pattern is used instead. The finished wafers will thus have three to five die locations replaced with test structures to monitor the process. Finally, copies are made from the master mask plate for actual use in the wafer processing area.

The mask layers are inspected to ensure they stack (align) on top of one another, in the same order as in actual use. A sample of various die locations on the mask is also inspected visually by trained personnel for defects in the pattern. If the number of defects is found to exceed the limit set by desired quality level and sample size, then the whole mask plate is rejected. For

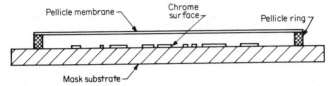

Figure 13-9 Pellicle covering for a mask to eliminate effects of particles. *(After Ref. 5. Reprinted with permission of Technical Publishing, a Company of Dunn & Bradstreet.)*

direct step-on wafer (DSW) aligners, the 10X reticles *are* the working plates. In this case, inspection is much more rigorous because a defect on the reticle would be replicated on every die.

Masks that are to be used in production are sometimes covered with a pellicle membrane.[5] Figure 13-9 shows a pellicle system schematic where the thin, high transmittance membrane is tautly mounted on a ring of suitable thickness. The ring is in turn mounted on the image plane of the mask. Dust particles, instead of collecting on the mask plate, collect on the surface of the pellicle. Because of the large stand-off distance, the particles are out of focus and would not print. Substantial savings in cost accrue from eliminating the usual mask cleaning steps and from an increase in yield.

13-3 AUTOMATED CHIP LAYOUT

The need to mechanize (computerize) the chip layout process is becoming more urgent. One factor is the shortage of good circuit designers and mask layout designers. Another is the fact that the layout process simply takes too long. Complex chips that require three to six mask layout designers from 6 months to 1 year to lay out and check out are not all that uncommon. A third factor is that an error in layout is one error too many if it causes the chip to be only partly functional. It requires corrections on the layout, a new set of mask plates to be ordered, and a new run (lot) to go through wafer fabrication. This entails excessive additional cost and delays product introduction. The solution lies in automating most if not all of the layout process.

The need for faster and error-free automated layout is critical as circuits enter the VLSI (very large scale integration) category. Technically, VLSI circuits are those containing more than 100,000 logic gates, but there is a connotation that the layout has to be highly mechanized. Otherwise, completion of the layout would simply be beyond reach. Of course several computerized layout aids have been developed and are continually being refined. Some were described earlier in Sec. 13-2. But since 1980, efforts have been directed at developing the mechanized design and layout activity into a cohesive discipline. The first step was to develop a set of generalized IC layout rules that are independent of particular fabrication sequence and divorced from the minutiae of circuit design. To that end, Mead and Conway and others[3,6,7] introduced the concept of lambda (λ)-based design rules. Lambda is an elementary distance unit defined from the assumption that the minimum feature size and minimum feature separation are each equal to 2λ. Furthermore, the worst-case misalignment between any two levels is λ, while the maximum edge shift due to processing is $\lambda/2$. Design rules are simpler: all features fall on a "lambda grid" and thus are much more suitable for novice designers. Furthermore, by simply scaling λ as needed, the design rules are highly portable from process to process, and remain applicable as a given process goes through shrink cycles. Such a concept works as long as technology makes nearly uniform advances, i.e., misalignments and smallest features on all mask levels scale down at roughly the same pace. It is understood that the inevitable loss in packing density from such simplifications is outweighed by the faster design and layout throughput. Proof that the above concept is commercially viable lies in the existence of "silicon foundries," companies able and willing to fabricate devices using other people's mask sets.

Most companies prefer to increase their layout efficiency not from design rule simplification, but from developing tools for automated chip layout. Their efforts have followed four basic methods[8] illustrated in Fig. 13-10.

Standard Cell

A large selection of logic building blocks such as counters and flip flops are laid out, fully characterized and stored in the layout system library. The designer calls for and places the needed cells and provides the interconnection. This is the most straightforward approach of the four. A more sophisticated algorithm can not only provide the routing of interconnects but can actually place the cells in such a way as to minimize lengths of interconnects. As Fig. 13-10 shows, standard cells have a uniform height, and interconnects are routed through wiring channels. Area utilization is rather poor since the wiring channels have to be high enough to accommodate the largest number of interconnects at any point along the channel.

Gate Arrays

The gate array is a variant of the standard cell layout that takes advantage of the economics of producing a semicustom chip instead of full custom. Also known as master slice, uncommitted logic array, and macrocell, it consists

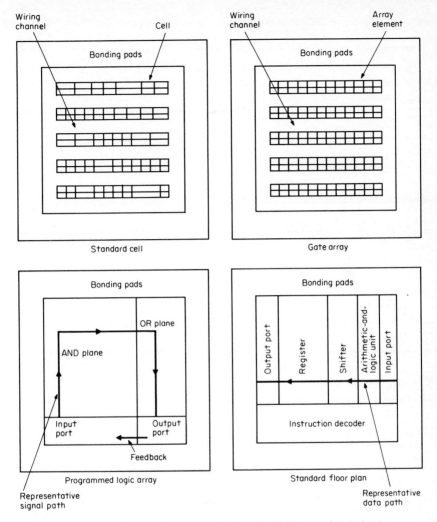

Figure 13-10 Four basic methods of automated chip layout *(After Ref. 8.)*

of a fixed array of identical logic circuits, as Fig. 13-10 shows. Around the periphery of the chip is circuitry for buffers and drivers. The key point is that all these logic building blocks are fixed in number and location. Designing a chip then consists of specifying the final wiring and interconnects.

Several economic advantages are derived from the gate array approach. Since all the chips are identical except at the last processing step, the cost of a new mask and separate production run is avoided. The chips can also be processed in advance and kept in inventory until the last step, shortening the time from product specification to delivery from several months to 5 to 10 weeks. The approach is so economically attractive that many companies

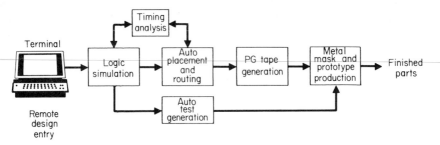

Figure 13-11 Flowchart for implementing gate array chip. *(After Ref. 9.)*

have offered the service of providing gate array chips.[9] In the early 1980s, these companies are offered chips with 500 to 2000 gates, but the complexity was expected to grow to 5000 to 10,000 range by the mid-1980s. Gate arrays are ideal for circuits with requirements of between 5000 to 100,000 units per year. Smaller quantity requirements can be met with EPROMs, while larger quantity requirements should consider a custom design.

MOS gate arrays are mostly implemented in CMOS. CMOS has greater design margin and thus simplifies the design effort. A two-layer metallization also facilitates layout by having interconnections within a logic block in the first metal layer and the interconnection between blocks on the second layer. The second layer is then allowed to run over the logic blocks.

Most users of outside gate array services usually do not have the full array of computer-aided design (CAD) tools available. Some may not even have prior circuit design experiences. Gate array service providers then must also supply training on readily accessible CAD tools. Figure 13-11 shows the sequence of events one must go through to produce a gate array chip. Circuit description is entered through a remote design entry terminal. Logic simulation, autoplacement, and routing results are used in timing analysis. Test vectors are also generated. Finally, the metal mask is produced by pattern generation, and fabrication of chips can begin.

The disadvantages of gate arrays are all associated with the layout patterns being fixed. The selection of logic blocks, the number of input and output buffers, and the amount of wiring area needed are based on probability calculations. To the extent that the estimates do not match each application perfectly, not all available logic gates are connected. Only 70 to 80 percent utilization is common. For this reason, gate arrays are provided in several sizes. Like standard cells, a gate array is a poor utilizer of chip area.

Programmed Logic Arrays

PLAs consist of NAND and NOR gate arrays that in series perform any Boolean logic operation [see Fig. 13-10 and also Sec. 9-5, "ROM (Read-Only Memory)"]. Algorithms are available to generate PLA layout directly from state machine equations, logic equations, or microcode specifications. PLA layout systems attempt to reduce chip size by logic minimization through reduction in the number of minterms. The regular form of PLA results in compact chip size and avoids excessively long connections.

Standard Floor Plan

The layout of a microprocessor generally follows a standard floor plan, analogous to the floor plan of a house with empty rooms to be filled as the user sees fit (see Fig. 13-10). The floor plan is efficient because it closely follows the data flow. Layout automation consists of trading off width for length within each core element of the plan. Cell heights can be stretched to match the continuous flow of the data path.

No one method of automating chip layout is optimum for all circuits. Standard cell and gate array is more appropriate for random logic, with the latter optimized for fast turnaround. PLAs are more suitable for controllers and microcode decoders. A standard floor plan is better for constructing the microprocessor data path—the calculation portion, which includes registers, shifters, and arithmetic logic units (ALUs) that share a common bus. Combinations of all methods are perhaps the proper approach toward an all-automated layout. The ultimate goal in CAD tools is the "silicon compiler"[10], a program that produces an IC layout from a functional description of a chip. The term compiler comes from the analogy with a software compiler that decomposes high-level language input into low-level machine codes.

13-4 ASSEMBLY

This section discusses the assembly and testing activities that transform completed wafers into final parts for sale.

Assembly Process

After wafers are fabricated, they are tested in wafer form. An operator lines up a probe card containing probe needles to match up to the chip's bond pads. If a particular die fails the test, it is marked with an ink dot. Good dies are not inked. After testing each die, the probe station automatically indexes to the next die location. The stepping distances are preset during machine setup. The machine steps along a row and moves up a row when the edge of the wafer is sensed.

The tested (and inked) wafers are sent to an assembly house for packaging. The assembly process begins by scribing (i.e., scoring) the "scribe alley" with a diamond tip. The scribe alley is a blank region (with exposed silicon) between chips allocated for this purpose. Scribing can be replaced with a sawing operation that cuts a kerf about a third through the wafer. With the wafer placed between cushioning papers, a rubber roller is passed over the wafer to break and separate the chips along the scored lines. The good, uninked dies are hand-picked with a vacuum pickup tool to continue with the assembly process.

The die is placed inside the package cavity in which has been placed a thin disk-shaped preform. When the package is heated to the gold-silicon eutectic temperature and the preform melts, the die is "scrubbed" to ensure a good bond to the package.

The next step is to provide wiring from the bond pad on the chip to the corresponding pattern on the package. If aluminum wires are used, they are ultrasonically bonded to form a wedge bond. If the wires are gold instead, thermocompression is used, leaving a compressed bead of wire at the bond site. The wiring can be either manual or automatic. The package is then ready for encapsulation.

In a ceramic package (see next section), a metallic lid is placed over the cavity and the package placed on a continuous-belt oven. Solder on the edge of the lid melts and effects a hermetic seal with the package. For a plastic package, the frame to which the die is bonded and wired is transfer-molded with thermosetting plastic. The leads of the plastic package are then bent and trimmed, and tinned for rigidity and good solderability.

The packages are packed in shipping sticks and sent to the final test. This final test is very exhaustive to guarantee meeting all specifications. It is usually performed at top speed and sometimes at more than one temperature. This test also catches any failure caused by assembly. Any yield loss at this step wastes not just the die but also the cost of assembly. Therefore, the earlier testing at the wafer level should ideally be stringent enough so as not to have fallout at the final test other than for assembly loss. However, there may be certain tests, such as maximum speed testing, that cannot be implemented at the wafer test level. The optimum target yield at the final test depends on trade-offs between the cost of the die, package, and testing, but would normally be in the 75 to 95 percent range.

Package Types

Package types are as myriad as there are special application requirements. The more common types used for MOS LSI ICs are plastic, ceramic [which includes side-brazed types and CERDIPs (ceramic dual in-line packages)], and leadless chip carriers (LCCs).

The plastic dual in-line package (DIP), shown in Fig. 13-12, has leads on two rows spaced 300 mils apart.[11] Each lead is separated 100 mils from its neighbor. Pin 1 is always the first pin going counterclockwise from a notch at one end of the package (top view). As the number of pins increases to 48, the width of the package takes incremental jumps to 400 then 600

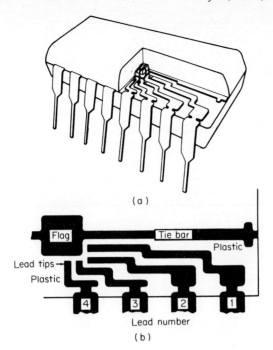

(a)

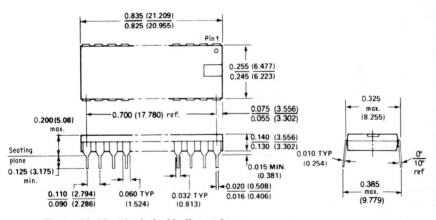

Figure 13-12 Plastic dual in-line package.

mils. During assembly, the die is attached to a tab or flag on the center of a metal frame. Leads are bonded from the die to fingers on the frame. The whole assembly is transfer-molded with thermosetting plastic, 12 to 24 DIPs per shot. Afterward, the external leads are trimmed to separate the DIPs from the frame. Plastic encapsulation is the lowest cost packaging technique.

Saw and
break

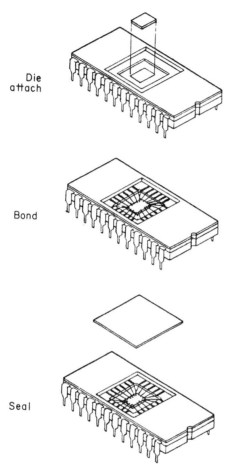

Die
attach

Bond

Seal

Figure 13-13 Side-brazed ceramic package *(Intel Corp.)*

However, it is not totally impervious to moisture, preventing its use in high-reliability applications unless the die itself is specially passivated.

Ceramic packages provide the capability to seal the chip hermetically in a cavity, and they come in two major forms. The first has package lead patterns side-brazed (fired) onto the ceramic package. Individual package leads are mechanically anchored to the sides while connected to these patterns. As shown in Fig. 13-13, the die is dropped into the cavity, the leads are bonded, and the lid sealed. The second form of ceramic package is the

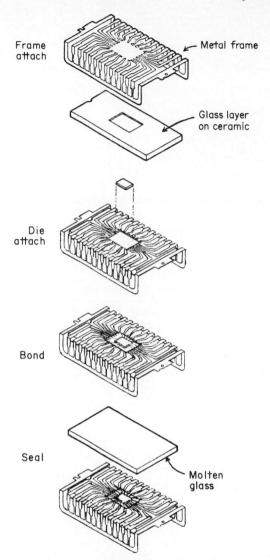

Figure 13-14 **CERDIP** package. *(Intel Corp.)*

CERDIP. As Fig. 13-14 shows, the chip and the surrounding frame are sandwiched between two ceramic plates. Bonding is provided by sealing glass that melts during the sealing operation.

Hermetic sealing protects the die from moisture and ionic contaminations. However, the cavity itself is not in vacuum. In fact, CERDIPs are usually sealed in air. Side-brazed ceramic packages are sealed in forming gas, a mixture of nitrogen and hydrogen.

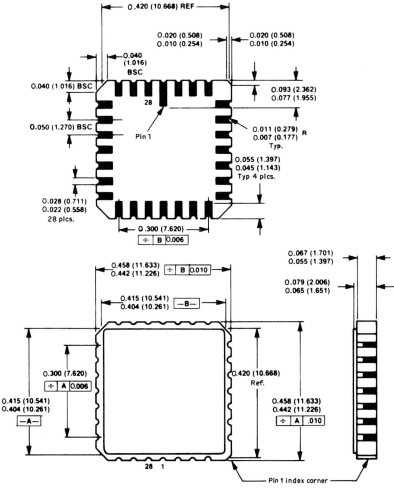

Figure 13-15 Ceramic leadless chip carrier package. All dimensions in inches (millimeters). A 28-leadless hermetic chip carrier (JEDEC Package Type C). (*After Ref. 13.*)

The leadless chip carrier[12,13] (LCC) is a ceramic package with leads on four sides that are spaced 50 mil apart (Fig. 13-15). The package lead patterns wrap around the sides and allow reflow soldering with matching pads on printed circuit boards. Replacement of bad packages is thus facilitated. The biggest advantage of LCCs, however, is the reduction in package area and its attendant weight savings. A reduction of up to 5:1 is possible. The Joint Electron Device Engineering Council (JEDEC) has set forth a line of standardized LCC package outlines. This has enhanced their acceptance and proliferation.

REFERENCES

1. R. H. Crawford, *MOSFET in Circuit Design,* Chap. 1, McGraw-Hill, New York, 1967.

2. American Microsystems, Inc., "Designing with gate arrays, Part 1: Technology and circuit elements," VLSI Design, special advertising supplement, May–June 1982.

3. C. Mead and L. Conway, *Introduction to VLSI Systems,* Chaps. 4 and 5, Addison-Wesley, Reading, Mass., 1980.

4. C. Machover and R. E. Blauth (eds.), *The CAD/CAM Handbook,* Chap. 3, Computervision Corp., Bedford, Mass., 1980.

5. R. Winn and R. Turnager, "Pellicles—an industry overview," *Solid-State Tech.,* pp. 41–43, June 1982.

6. C. H. Séquin, "Generalized IC layout rules and layout representations," in *VLSI 81,* edited by J. P. Gray, Academic, pp. 13–23, 1981.

7. M. Marshall et al., "The 1981 achievement award," *Electronics,* pp. 102–105, Oct. 20, 1981.

8. S. Trimberger, "Automating chip layout," *IEEE Spectrum,* pp. 38–45, June 1982.

9. B. Groves, "Do it yourself VLSI," *Digital Design,* pp. 60–73, September 1982.

10. J. Werner, "The silicon compiler: Panacea, wishful thinking, or old hay?" *VLSI Design,* pp. 46–52, September–October 1982.

11. Intel, *Component Data Catalog 1982,* Chap. 14, Intel Corp., Santa Clara, Calif., 1982.

12. J. A. Bauer, "Chip carrier packaging applications," *IEEE Trans. Components, Hybrids, Manufacturing Technology,* vol. CHMT-3, no. 1, pp. 120–125, March 1980.

13. D. Yeskey, "Leadless chip carrier," *Intel Technology Report,* Tr-1, June 1982.

14

Yield and Reliability

14-1 IMPORTANCE OF YIELD

Yield is that fraction of the total die locations on a wafer that results in working circuits. It is of utmost importance to both wafer processing engineers and circuit designers. In a wafer processing area, the cost of processing a wafer is relatively constant for a given technology. The cost per *good* die is then completely dependent on how many good dies the wafer yields. The improvement in yield as the processing area "goes down the learning curve" is the main reason integrated circuits have the enviable trait of constant (and sometimes dramatic) price reductions. The chip designer is very much concerned about yield because the technology picked strongly determines the cost per wafer; therefore the only way the cost of the chip will go down is, again, to increase the yield. We shall see shortly that the chip designer has dramatic control over this by the size of the chip. If he or she is designing a large system, she or he has an extra degree of freedom in lowering the total cost of the system by partitioning it into several chips. Figure 14-1 shows that the cost of a die goes down significantly as the chip size is decreased, a fact we will demonstrate later in the chapter. Therefore one stands to gain by making a set of very small chips rather than one large one. However, the cost of handling and testing a larger number of chips, and certainly the cost of providing a larger number of packages increases with more chips in the system. But one will often find that the minimum cost is obtained by partitioning the system into more than one chip. It should be pointed out, however, that the drive to integrate more and more circuit functions into one chip is relentless because for many circuits, the cost of testing and packaging is easily one to two times the cost of the die itself.

So how does one predict the yield of an integrated circuit? Yield sometimes fluctuates over a wide range and changes over a period of time. But if

339

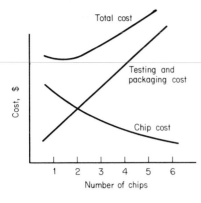

Figure 14-1 System partitioning to achieve minimum overall cost.

a circuit is designed and tested properly, and is processed in an area that runs a process with some consistency, its yield will closely follow certain well-behaved statistical laws (as long as the yield is determined from a large enough statistical sample).

14-2 YIELD MODELS

Let us assume that a wafer has a certain number of killing defects (defects that cause circuit malfunction) spread throughout the wafer. One can then define an average defect density D_0. If a chip has an area A, then it has, on the average, AD_0 number of defects per chip.

If the number of defects is a uniformly distributed random variable, then the probability of finding k number of defects inside one single die is given by the Poisson distribution (see Fig. 14-2).

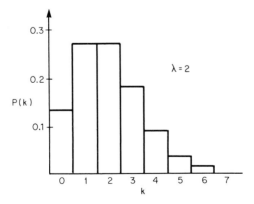

Figure 14-2 Poisson probability distribution function.

(14-1) $\quad P(k) = \dfrac{\lambda^k e^{-\lambda}}{k!} \qquad k = 0, 1, 2, 3, \ldots$

where

(14-2) $\quad \lambda = AD_0$

Therefore

(14-3) $\quad P(k) = \dfrac{(AD_0)^k e^{-AD_0}}{k!} \qquad k = 0, 1, 2, 3, \ldots$

The yield is the probability of finding a chip with zero defect, i.e., yield is

(14-4) $\quad Y = P(k = 0) = e^{-AD_0}$

The yield equation given by Eq. (14-4) is the Poisson yield model. That model is found to be too conservative, that is, it predicts too low a yield, especially for large devices (e.g., $AD_0 > 1$). One must take into account that the defect density is not uniform and varies from wafer to wafer, since they all cannot be handled exactly the same way. And even within a wafer, the defect density is not constant. It is known to be higher at the edge of the wafer than at the center. Or the defects may be clustered due to point or line defects. This requires that we modify our yield model to include a defect density distribution, so that yield is now

(14-5) $\quad Y = \displaystyle\int_0^\infty F(D)e^{-AD}\, dD$

The various forms of the defect density distribution $F(D)$ assumed give rise to several yield formulas.[1]

The Poisson model mentioned earlier is a special case of Eq. (14-5) where $F(D)$ is a delta function at $D = D_0$ (see Fig. 14-3).

Murphy[2] assumed in his yield model that $F(D)$ is gaussian, but he approximated it with a triangular function shown in Fig. 14-4 and described below:

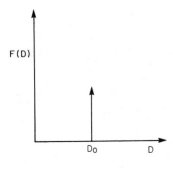

Figure 14-3 Defect density distribution in the form of a delta function.

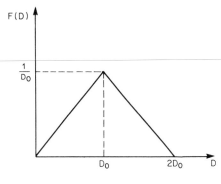

Figure 14-4 Defect density distribution in the form of a triangle, used by Murphy. (*After Ref. 2.*)

$$
(14\text{-}6) \quad F(D) = \begin{cases} \dfrac{D}{D_0^2} & 0 \le D \le D_0 \\[2ex] \dfrac{2}{D_0} - \dfrac{D}{D_0^2} & D_0 \le D \end{cases}
$$

Substituting Eq. (14-6) into Eq. (14-5) results in Murphy's yield model:

$$
(14\text{-}7) \quad Y = \left(\frac{1 - e^{-AD_0}}{AD_0} \right)^2
$$

The Seeds model,[3] on the other hand, is more optimistic in that it assumes low-defect density to be more likely than high-defect density, leading to an exponential defect density distribution (see Fig. 14-5):

$$
(14\text{-}8) \quad F(D) = \frac{1}{D_0} e^{-D/D_0}
$$

This results in the Seeds model of

$$
(14\text{-}9) \quad Y = \exp\left(-\sqrt{AD_0}\right)
$$

A fourth model is the Bose-Einstein model which uses Bose-Einstein statistics to predict yield. This model assumes that most defects are masking-related and that for n masking steps and an average D_0 defect density, the yield is

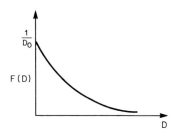

Figure 14-5 Defect density distribution in the form of an exponential function.

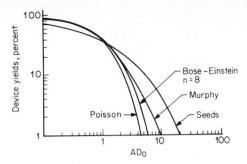

Figure 14-6 Comparison of various yield models. *(After Ref. 1.)*

(14-10) $\quad Y = \dfrac{1}{(1 + AD_0)^n}$

The Price model[4] is a special case of Eq. (14-10) with $n = 1$. One criticism of the Bose-Einstein model is that although it does take into account the number of masking layers, it incorrectly assigns equal weight to each layer and uses one single D_0 value. For example, an ion implantation step does not generate nearly as many defects as the metal masking step.

A comparison of the various yield models is seen graphically in Fig. 14-6. Their differences are more pronounced when AD_0 is large.

A more useful way of plotting a yield model is exemplified by Fig. 14-7 where the Murphy model is displayed by plotting yield as a function of area with the defect density D_0 as a parameter. D_0 is most commonly expressed in defects per square centimeter. Figure 14-7 allows one to visualize the impact of changing the die size assuming a constant average defect density D_0. Recall that D_0 can safely be assumed a constant for a given technology, for a given fabrication line, and within a certain period of time.

A word of caution is needed at this point concerning what to expect in using yield models. It is not the intent of using yield models to come up with a defect density D_0 that will allow one to count the defects on a chip *visually*. It is rather a statistical parameter to describe the electrical yield behavior of a large number of chips. It should be kept in mind that even the yield of a circuit is difficult to define precisely. One can imagine that for a chip of a given size, the yield will depend on how densely the whole chip is filled with active circuitry, how much design margin is incorporated, and how tight the test limits are.

Knowing the yield of a wafer, one would need to know the number of available die locations in order to obtain the actual number of good dies per wafer. The gross die on a wafer is given by

(14-11) $\quad$ Gross die $= \dfrac{\pi(R - A^{1/2})^2}{A}$

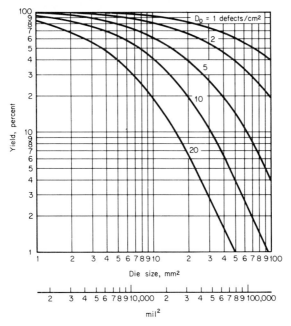

Figure 14-7 Murphy model yield curve.

where R is the wafer radius and A is the die area. Gross die as a function of die area for various wafer diameters is plotted in Fig. 14-8. When the gross die is multiplied by the yield, one would obtain a figure such as Fig. 14-9 showing the total number of good dies expected as a function of die size. That figure emphasizes the point brought up earlier in the chapter on how sharply the yield drops for large die size.

14-3 RELIABILITY AND FAILURE MODELS

The use of integrated circuits over discrete transistors increases the reliability of a system. It stands to reason that the reliability of integrated circuits should be studied in order to uphold that improved reliability. Indeed it will be demonstrated later in the chapter that the failure rate as seen by end users can be improved by techniques such as burn-in to screen out initial defects and to move devices to a more favorable region of their life cycle.

Reliability should first be designed into the integrated circuit. The process should be developed and manufactured to be inherently reliable and not to be prone to early failure. Some of the failure mechanisms that are known and should be designed against include electromigration in metal, ionic contamination causing threshold drift, passivation layers not completely stopping contamination, and corrosion.[5] Similarly, the circuit design should provide a large design margin to counteract inevitable degradation.

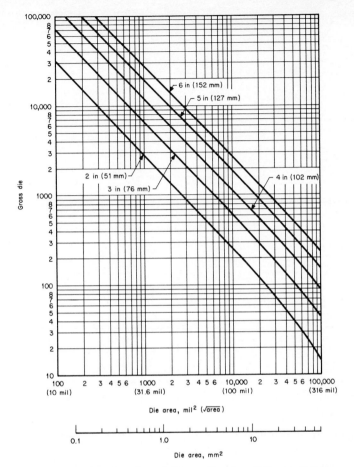

Figure 14-8 **Gross number of dies as a function of die area for various wafer diameters. Gross die** $= \pi (R - A^{1/2})^2/A$; R = **wafer radius;** A = **die area.** *(After Ref. 11.)*

Once a part is designed to be technically and economically as reliable as possible, the residual failure rates can be quantified by statistical models. Failure can be narrowly defined as nonconformance to performance specification or loosely defined as complete nonfunctionality. In any case, whichever criterion is initially chosen should be consistently applied throughout.

Failure mode is the electrical cause of the rejection, e.g., slow access time, which will point toward the basic physical *failure mechanism,* e.g., ionic contamination causing high V_T. The failure of a part is a random event with the random variable being the time to failure. This random variable is found to fit a lognormal distribution for semiconductor devices. Such a distribution also fits rate-dependent exponential processes such as chemical reactions and diffusions. The lognormal distribution is shown in Fig. 14-10 where the

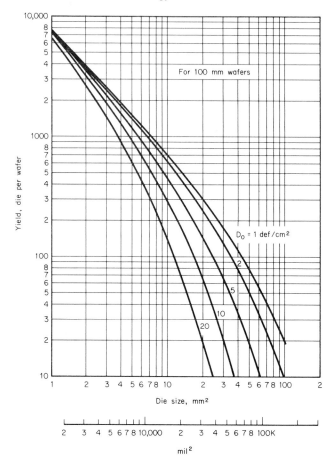

Figure 14-9 Total number of dies yielded as a function of die area for 100-mm-diameter wafers using Murphy's yield model.

logarithm of the time to failure t has a gaussian (normal) distribution.[6] Mathematically, a normal distribution has a probability density function expressed as

(14-12) $$f(x) = \frac{1}{\sqrt{2\pi}\,\sigma} \exp -\frac{z^2}{2}$$

(14-13) where $x = \log t$
$z = [x - \mu]/\sigma$
$\mu = $ mean of $\log t$
$\sigma = $ standard deviation of $\log t$

The cumulative distribution function is the integral of the probability density function from $-\infty$ to x:

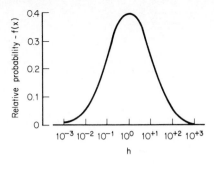

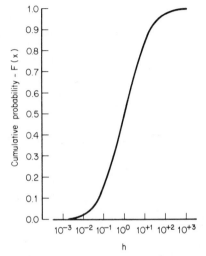

Figure 14-10 The probability density function and cumulative distribution function for the lognormal distribution.

$$(14\text{-}14) \quad F(x) = \frac{1}{\sqrt{2\pi}\,\sigma} \int_{-\infty}^{x} \exp{-\frac{z^2}{2}}\, dz$$

That function, shown also in Fig. 14-10, is the probability of having a time to failure less than or equal to x, i.e., the cumulative failure. With an exchange of the X and Y axes, and an appropriate change in variable, a chart can be drawn such that the cumulative distribution function for a lognormal distribution results in a straight line, as Fig. 14-11 illustrates. Two parameters completely specify each distribution (each line). The mean is the point on the log t axis with 50 percent cumulative failures. That mean is the median life and is designated t_{50}. The standard deviation, or sigma, is given by

$$(14\text{-}15) \quad \sigma = \ln t_{50} - \ln t_{16} = \ln \frac{t_{50}}{t_{16}}$$

where t_{16} is the time to 16 percent cumulative failure. In Fig. 14-11a are plotted two distributions with a median life of 100 h and sigma of 1 and 2 (dimensionless). Figure 14-11b is the same graph for the reader's use.

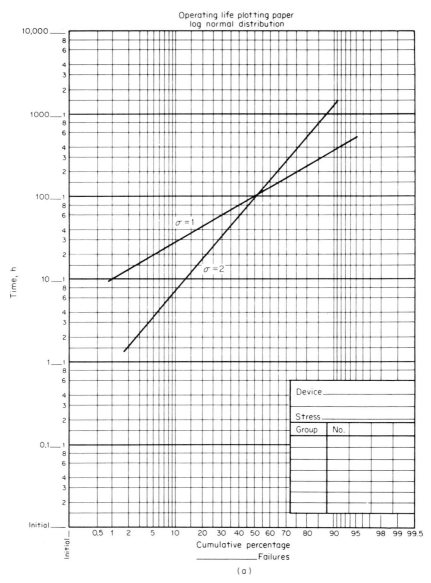

Figure 14-11 Probability graph where a lognormal cumulative distribution is plotted as a straight line (*a*) with sigma = 1 and 2 plotted; (*b*) for reader's use. (*After Ref. 13.*)

To determine mean and standard deviation for a particular part, one would perform a life test. A life test implies the application of accelerating stress such as high temperature and/or high voltage. Otherwise the test would take too long to conduct. A life test also implies working with a sample of the main population. To be accurate, the life test should last until at

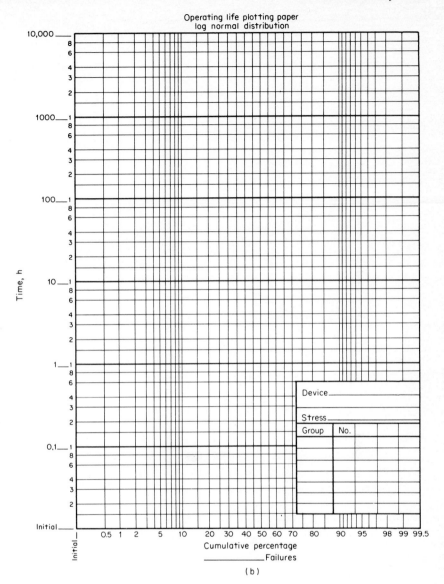

Figure 14-11 (continued)

least 50 percent of the samples fail. When failures are plotted on a lognormal probability graph, the slope of the graph gives the standard deviation. A low standard deviation (<2) yields a near-horizontal line and is an indication of a well-controlled process. A high standard deviation ($\gg 2$) yields a more vertical line and indicates a process out of control. If two life tests are performed at different temperatures, the two resulting lines should be par-

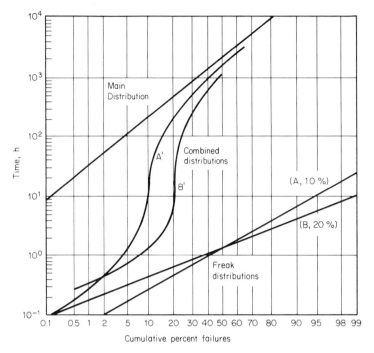

Figure 14-12 An S-shaped graph is an indication of bimodal distribution with early failure (infant mortality). *(After Ref. 7.)*

allel for the same failure mechanism. Otherwise, for a certain part of the population, higher stress will result in longer life, which is contrary to intuition and experience.

A life test result would commonly show not a straight line but an S-shaped curve[7] such as in Fig. 14-12. It indicates a bimodal distribution with a small fraction of the population distinctly failing earlier. This high initial failure rate is referred to as infant mortality, or freak distribution, and shows up in the more familar "bathtub" curve (Fig. 14-13). The failure rate is initially high because of infant mortality, then drops to a relatively

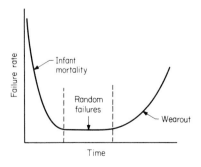

Figure 14-13 "Bathtub" failure rate curve.

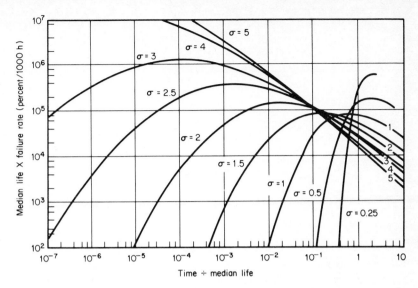

Figure 14-14 Instantaneous failure rate for lognormal distribution. *(From L. R. Goldthwaite, Bell Telephone Monograph 3814.)*

constant rate that is representative of the main population, and finally increases again due to wearout mechanisms. However, wearout is not pronounced in electronic circuits. Figure 14-12 shows the resulting distribution when a 10 or 20 percent freak population is present. The inflection in the curve is the point in time when the freak population has all failed.

The instantaneous failure rate[8] at a given point in time is the probability density function Eq. (14-12) expressed as a function of time $f(t)$, divided by the probability of surviving up to time t, i.e., $1 - F(t)$. For the lognormal distribution, results are presented in the Goldthwaite[9] graph of Fig. 14-14. Both the abscissa and ordinate are normalized to median life t_{50}. The instantaneous failure rate, rather than being constant, peaks generally before median life and constantly decreases thereafter.

A unit of measure of failure rate used in telephony and other high-reliability areas is one failure unit (FIT). One FIT is one failure in 10^9 device-hours. Therefore, 1000 FITs have 0.1 percent failure per 1000 h. MOS ICs typically have a reliability of a few hundred FITs.

In a system with several components, the overall failure rate is the sum of the individual failure rates. Since the failure rate per component has been kept relatively constant, system reliability is enhanced if a higher level of integration is achieved that results in a fewer parts count.

14-4 BURN-IN

For most failure mechanisms, life testing of semiconductor components can be drastically accelerated by operation at elevated temperatures. Testing results can then be extrapolated or derated back to normal operation. High-temperature burn-in also culls out potential infant mortality and moves devices to regions of lower failure rate.

The reaction rate of chemical processes is related to temperature according to the Arrhenius equation:

$$(14\text{-}16) \quad R(T) = C \exp \frac{-E_a}{kT}$$

where R = reaction rate
C = constant
E_a = activation energy, eV
k = Boltzmann's constant (8.625×10^{-5} eV/K)
T = temperature, K

Thermodynamic theory states that reactant molecules need to exceed a minimum activation energy before reaction will take place. The fraction of reactants that have energies which exceed E_a is given by the exponential term.

It is found that the time to failure also has a temperature dependence that follows the Arrhenius equation, Eq. (14-16). One can find the activation energy for a given failure mechanism by taking the natural logarithm of both sides which yields

$$(14\text{-}17) \quad \ln t = -\frac{E_a}{kT} + C$$

E_a then is the magnitude of the slope when $\ln t$ is plotted against $1/T$. That is

$$(14\text{-}18) \quad E_a = \frac{k \ln (t_1/t_2)}{1/T_2 - 1/T_1}$$

The failure points t_1 and t_2 should be measured at the same distribution point, e.g., median life t_{50}. The temperature should be the junction temperature, and not just ambient or case temperature. Figure 14-15 shows a special graph that facilitates the use of Eq. (14-18). One axis is in logarithm of time while the other is in linear $1/T$ scale. The slope of a straight line would correspond to the activation energy. The activation energy can be read off the vertical scale on the right by drawing a line with similar slope to pass through the "bull's-eye" at the top of the graph. A line for 1.0 eV is shown as an example. Typical activation energies[10-12] are shown in Table 14-1. A large activation energy corresponds to a large acceleration factor from low to high temperature.

The application of accelerated testing to predict the failure rate is summarized in the procedure below:

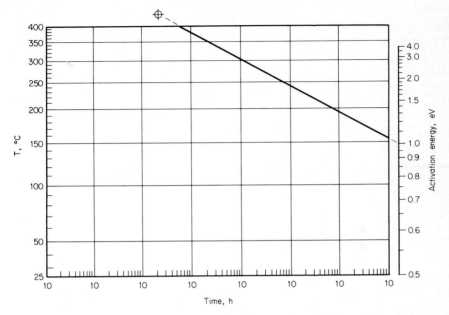

Figure 14-15 Graph for plotting Arrhenius equations and determining activation energy. *(After Ref. 11.)*

1. Perform an accelerated testing at a given elevated temperature until at least 50 percent of the samples fail. Analyze the failed parts to ensure they are all of one dominant failure mode. If they are not, the whole procedure may have to be repeated for each failure mode.

2. Plot the failures vs. time on the lognormal cumulative distribution graph (Fig. 14-11). Draw a straight line through the data for a best fit. Parameters such as median life t_{50}, and sigma, σ, can be determined from the straight line. An S-shaped curve indicates the presence of infant mortality.

3. Repeat the above steps for two other elevated temperatures. The resulting lines should be parallel, indicating an acceleration in time to failure. A shift in sigma would indicate a different failure mechanism coming into play.

4. Transfer the data at the identical cumulative failure point, e.g., t_{50}, to an Arrhenius graph (Fig. 14-15). Other failure points, e.g., t_{10} or t_{20}, should yield parallel lines unless corrupting factors such as infant mortality are present. Determine the activation energy, E_a.

TABLE 14-1 Experimentally Observed Activation Energy

Failure mechanism	E_a, eV
Ionic contamination	1.0–1.4
Oxide breakdown	0.3–0.35
Charge injection (slow trapping)	1.0–1.3
Electromigration	0.5–1.2
Gold-aluminum bond failure ("purple plague")	1.0–1.05
Corrosion, electrolytic	0.3–0.6
Degradation in 80% relative humidity	0.9
Polysilicon and silicon defects	0.5
Charge loss and gain (EPROMs)	0.6

Source: After Refs. 10, 11, and 12.

5. Compare the obtained E_a with those observed by others (Table 14-1) to confirm the suspected failure mechanism and for overall reasonableness of data.

6. Extrapolate to the highest rated normal temperature and determine t_{50}. With shape parameters consisting of t_{50} and σ, instantaneous failure rate at any point in the operating life can be determined by the use of Fig. 14-14.

REFERENCES

1. M. R. Gulett, "A practical method of predicting IC yields," *Semiconductor Int.*, pp. 87–94, March 1981.

2. B. T. Murphy, "Cost-size optima of monolithic integrated circuits," *Proc. IEEE*, pp. 1537–1545, December 1964.

3. R. B. Seeds, "Yield and cost analysis of bipolar LSI," *IEEE Int. Electr. Dev. Meeting*, Washington, D.C., p. 12, October 1967.

4. J. E. Price, "A new look at yield of integrated circuits," *Proc. IEEE*, pp. 1290–1291, August 1970.

5. M. J. Howes and D. V. Morgan, *Reliability and Degradation: Semiconductor Devices and Circuits*, Chap. 4, Wiley, 1981.

6. J. Eachus, "Failure analysis in brief," *Semiconductor Int.*, pp. 103–112, January 1982.

7. D. S. Peck, "The analysis of data from accelerated stress tests," *9th Ann. Proc. Rel. Phys.*, pp. 69–78, 1971.

8. A. B. Glaser and G. E. Subak-Sharpe, *Integrated Circuit Engineering,* Chap. 16, Addison-Wesley, Reading, Mass., 1977.

9. L. R. Goldthwaite, "Failure rate study for the log-normal lifetime model," *Proc. 7th National Symposium on Reliability and Quality Control,* pp. 208–213, January 1961.

10. Intel, "Reliability monitor program," Available from Intel literature department, 3065 Bowers Ave., Santa Clara, Calif. 95051, Order Number 210301-001, 1981.

11. O. D. Trapp et al., *Semiconductor Technology Handbook,* Chaps. 14 and 15, Technology Associates, Portola Valley, Calif., 1982. (*Note:* This handbook contains many charts and graphs that are useful in yield analysis.)

12. D. S. Peck and C. H. Zierdt, Jr., "The reliability of semiconductor devices in the Bell System," *Proc. IEEE,* vol. 62, pp. 185–211, February 1974.

13. D. S. Peck and O. D. Trapp, *Accelerated Testing Handbook,* Technology Associates, Portola Valley, Calif., 1981. (These graphic analysis forms are available from Technology Associates, 51 Hillbrook Drive, Portola Valley, Calif., 94025.)

PROBLEMS

1. A proposed circuit is estimated to require 40,000 mils^2 of chip area. It is to be fabricated on a 4-in-diameter wafer that costs $100 to process. If the process has a defect density of five defects per square centimeter, what will be the final cost if packaging, testing, and assembly yield loss is quantified at $1.50 per finished IC?

   It is proposed that the chip be partitioned into two. Assume that each of the partitioned chips requires a 10 percent increase in area for additional I/O buffers. What would be the total cost of the two-chip set if the other costs stay at $1.50 per finished IC?

2. What percentage gain in the number of dies is available by going from 3- to 4-in wafers? Compare that to the incremental change in going to 5- and to 6-in wafers.

3. A 4-in wafer has a chip with 20 mm^2 of area. The process is rated at one defect per square centimeter. A more advanced process is available that will allow a 30 percent linear shrink (0.7 ×). But because it's a newer process and much denser, it is estimated to run only at five defects per square centimeter. Does it make sense to switch to this new process, assuming a 20 percent increase in wafer cost?

   At what chip size will the break-even point be reached such that the cost will be the same for either process?

4. Given an S-shaped distribution as shown in Fig. 14-12, describe how one might extract the freak and main distribution lines from the single curve.

5. Three lots of 50 samples each are randomly pulled from the main population. They are subjected to operation at elevated temperatures of 225, 250, and 275°C, respectively. The parts themselves generate negligible power. At predetermined time

TABLE 14-2

Test time, hours	Lot A, 225°C	Lot B, 250°C	Lot C, 275°C
4	0	0	2
8	0	1	4
24	0	3	14
48	1	8	10
100	2	12	12
200	7	11	6
500	17	10	1

periods they are pulled from the burn-in ovens for testing. The measured failures are recorded in Table 14-2. Plot their distribution on lognormal probability paper. Are the three distributions parallel? Determine their standard deviation σ.

Determine the activation energy and the acceleration factor operating from 80 to 225°C. What is the failure rate at 80°C at 10 years of product life?

Physical Constants

Electronic charge	$q = 1.602 \times 10^{-19}$ coulomb
Electronvolt	$1 \text{ eV} = 1.602 \times 10^{-19}$ joule
Permittivity of free space	$\epsilon_o = 8.86 \times 10^{-14}$ F/cm
Boltzmann's constant	$k = 8.62 \times 10^{-5}$ eV/K
Thermal voltage (room temperature, 300 K)	$kT/q = 0.0259$ V

Material Properties

SILICON

Atomic number	$z = 14$
Atomic weight	$= 28.09$
Breakdown field	$E_B = {\sim}3 \times 10^5$ V/cm
Energy gap 300 K	$E_G = 1.124$ eV
0 K	$E_G = 1.170$ eV
Intrinsic carrier concentration	$n_i = 1.5 \times 10^{10}$ cm^{-3}
Lattice constant	$a_o = 5.43$ Å
Melting point	$T_m = 1412°C$
Mobility (low doping): Electron	$\mu_n = 1350$ cm^2/V · s
Holes	$\mu_p = 480$ cm^2/V · s
Relative permittivity (dielectric constant)	$\epsilon_s = 11.7$

INSULATORS

	SiO$_2$	Si$_3$N$_4$
Energy gap (300 K)	$E_G = 9$ eV	4.7 eV
Index of refraction	$n = 1.46$	2.0
Relative permittivity	$\epsilon_{ox} = 3.9$	7.5

Source: H. F. Wolf, *Semiconductors*, Wiley, New York, 1971; R. S. Muller and T. I. Kamins, *Device Electronics for Integrated Circuits*, Wiley, New York, 1977; S. M. Sze, *Physics of Semiconductor Devices*, Wiley, New York, 1981.

Conversion Constants

$$1 \text{ mil} = 0.001 \text{ in} = 25.4 \text{ } \mu m$$
$$1 \text{ } \mu m = 10^{-4} \text{ cm} = 10,000 \text{ Å}$$
$$1 \text{ Å} = 10^{-8} \text{ cm} = 10^{-4} \text{ } \mu m$$
$$1 \text{ cm} = 0.3937 \text{ in}$$
$$1 \text{ in} = 2.54 \text{ cm}$$
$$1 \text{ mm} = 39.37 \text{ mil}$$
$$1 \text{ mil} = 2.54 \times 10^{-2} \text{ mm}$$
$$1 \text{ mm}^2 = 1550 \text{ mil}^2$$
$$1000 \text{ mil}^2 = 0.645 \text{ mm}^2$$
$$1 \text{ h} = 3.6 \times 10^3 \text{ s}$$
$$1 \text{ day} = 8.64 \times 10^4 \text{ s}$$
$$1 \text{ week} = 168 \text{ h}$$
$$1 \text{ month} = 720 \text{ h}$$
$$1 \text{ year} = 8760 \text{ h}$$

Solutions to Selected Problems

Chapter 1

1. (a) Bilaterally symmetric means that device characteristics are not altered when source and drain are interchanged.
   (b) Unipolar means that only one carrier type is used for current conduction.
2. Electrons/positive/negative/electrons.
3. The equation given is the triode equation, which is not valid in saturation region.
4. Saturation region.

Chapter 2

1. $V_B = -(kT/q) \ln (N_D N_A / n_i^2)$.
2. True for ideal diode; false for silicon junction which is dominated by generation current.
3. Positive terminal on n-side.
5. (a) $p = 5 \times 10^{15}/\text{cm}^3$; (b) $n = 4.5 \times 10^4/\text{cm}^3$; (c) resistivity $= 2 \text{ } \Omega \cdot \text{cm}$.

Chapter 3

2. Surface concentration of $Q_n = N_A$ in the bulk.
3. (a) $V_{TN} = 0.23$ V; (b) $V_{TP} = -2.32$ V.
4. (a) $V_{TFP} = 22.6$ V; (b) $V_{TFM} = 33.9$ V.
5. $|dV/dT| = -0.0025$ V/°C (approx).
6. Change in $V_T = 1.1$ V.

Chapter 4

1. For both positive and negative ions, negative bias results in positive CV shift.
2. (a) Larger/larger; (b) decreases/increases.
3. False.
4. n-channel: $C_{FB} = 17.62$ pF, $V_T = 0.63$ V; p-channel: $C_{FB} = 16.30$ pF, $V_T = -0.31$ V.

Chapter 5

1. W/L is replaced by $2\pi/\ln(r_2/r_1)$.
3. $V_T = 0.9$ V; $\beta = 1.6 \times 10^{-6}$ A/V^2; V_T (10–40 method) $= 0.9$ V.
4. There is little change in V_T due to back-gate bias. Speed is improved because junction capacitance is reduced.
5. Current does not drop to zero, but remains at a fixed minimum value.

Chapter 6

1. The charge in the depletion region is contained in trapezoid $BFGC$ of Fig. 6-6.
2. The charge in the depletion region is represented by the trapezoid under the gate of Fig. 6-7. Solve for ΔL by using the triangle encompassed by ABC in the figure.
3. $V_T = -0.03$ V.

Chapter 7

1. $V_{TP} = -4.08$ V; $V_{TF} = -36.9$ V.
2. $V_{TN} = -0.32$ V (depletion device).
3. Aluminum melts at 660°C.
4. The contact enhancement prevents a short to substrate even with misaligned contact.

Chapter 8

1. Final oxide thickness $= 900$ Å.
2. Silicon step $= 1380$ Å.
3. (a) Predeposition junction depth $= 0.65$ μm; (b) drive-in junction depth $= 4.29$ μm.
5. Equivalent thickness $= 878$ Å; fraction in silicon $= 99.2\%$.
7. Etch time for deposited oxide $= 28.6$ s; etch time for thermal oxide $= 94.3$ s.

Chapter 9

1. Inverter gain ratio = 5.
2. For standard definition: $NM^1 = 3.4$ V; $NM^0 = 1.0$ V.
   For less conservative definition:
   $NM^1 = 3.6$ V; $NM^0 = 1.2$ V.
3. Worst case is when temperature is high and power supply is low.
4. For output rising, $t_{pd}^+ = 50$ ns. For output falling, $t_{pd}^- = 40$ ns.
6. It is a NOR gate.

Chapter 10

1. $\lambda = x_d/2(L - x_d)(V_D + V_B)$.
2. Figure 10-8b is a class B source-follower output stage, swings only to one threshold drop from supply rails, and has unity voltage gain and high output impedance. A resistive load further limits voltage swing.
   Figure 10.31 operates with high voltage gain and has higher output impedance, but is less affected by resistive load.
3. $dV_{out}/dV_{in} = 0.50$.
5. Gate to drain overlap capacitance would leave residual voltage error on holding capacitor. One solution is to inject opposite polarity error to the capacitor with a similar overlap capacitance.
7. $Q_{n,max} = C_{o,st}(\phi_{s,st} - \phi_{s,tr}) + qN_A(x_{d,st} - x_{d,tr})$.

Chapter 11

1. Equate saturation current through p- and n-channel devices.
2. Static power dissipation = 1 μW; dynamic power dissipation = 100 μW at 1 MHz.
3. $V_{in} = 3.31$ V for $V_{DD} = 10$ V; $V_{in} = 1.77$ V for $V_{DD} = 5$ V.
4. The modified n-channel device results in more constant total resistance as a function of input voltage.

Chapter 14

1. (a) Number of available die = 260; number of good die = 81; cost = $2.73.
   (b) Number of available die = 500; number of good die = 250; cost = $3.80. A single package is still lower in cost.
2. Gain in number of die is 78% from 3 in. to 4 in.; 44% from 4 in. to 5 in.
3. (a) Number of good die is 300 for old process; 440 for new process. Cost savings = 18%.
   (b) New yield = 120% of old yield.
5. $\sigma = t_{50}/t_{16} = 3$; $E_a = 1.2$ eV; acceleration factor = 10^5; median life = 5 × 10^7 h; failure rate = $1.6 × 10^{-2}$ % per 1000 hours.

Index

AC analysis sample run, 283–299
Acceptors, 18–19
Accumulation, 33
Activation energy, 352
Aliasing, 241
Alignment sequence, 320
Alpha particles, 210
Aluminum:
  deposition of, 174–175
  with 1% silicon, 175
  as self-aligned gate material, 147
  sputtering deposition of, 175
Amplifier gain:
  analysis of, 224–225
  with complementary load, 228–230
  with depletion load, 228
  with saturation load, 227–228
Analog building blocks, 227–233
Anneal:
  after oxidation, 43, 64
  aluminum deposition, 175
  ion implantation, 168–172
Arrhenius equation, 352
Arsenic-doped glass, 143
Arsenic source and drain, 143
ASPEC (Advanced Simulation Program for Electronic Circuits) circuit simulation program, 280–316
Assembly process, 331–332
Avalanche region, 86–88

Back-gate bias, 91–94
Band bending (*see* Energy-band diagram)
Band diagram (*see* Energy-band diagram)
Band gap (E_G), 144
Bathtub curve, 350
Bias-heat stress technique, 63
Bilaterally symmetric characteristic of MOS transistor, 5

Bipolar transistor, comparison of, with MOS, 7–8
Bird's beak, 139
Block erase cycle, 216
Blowbacks, 326
Bootstrap circuit, 218–219
Bose-Einstein yield model, 342
Breakdown voltage:
  of drain junction, 86–87
  of *pn* junction, 26–27
Buffered etch, 180
Built-in potential V_B, 21
Bulk charge, varying, 94–95
Buried contacts, 144
Burn-in, 351–352
Byte-wide ERASE and WRITE, 216

Capacitance:
  depletion-layer, 28–29
  device, 225–226
Capacitance-vs.-Time (*CT*) plot, 71–74
Capacitor-based circuits, 237–239
Ceramic package, side-brazed, 334–335
CERDIP package, 334–335
Channel length modulation (shortening), 85, 225
Channeling, 166
Channelstop, 248
Charge-coupled device (CCD):
  buried-channel device, 248
  dark current, 71, 251
  fill and spill technique, 252
  nonequilibrium condition, 58
  serial-parallel-serial (SPS) format, 253
  similarity with dynamic RAM, 210
  split-electrode technique, 254–255
  surface-channel device, 248
  surface potential of, 245–247
  tap weights, 254
  transfer inefficiency, 249

Checkout, 325–326
Chemical vapor deposition (CVD), 173–174
Circuit simulation program, need for, 279–280
Class 100 rating, 184
Clean, 182–184
CMOS (Complementary MOS):
  advantages and disadvantages of, 261–262
  circuit analysis, 265–266
  definition of, 5
  gate arrays, 330
  increased processing complexity, 268–270
  inverter, 262–263
  latch-up, 270–275
  logic gates, 263–264
  power dissipation, 266–268
  static RAM cell, 265
  transmission gates, 264
Compensation capacitor, 235
Compiler, silicon, 331
Complementary error function, 157
  (*See also* Diffusion)
Composite drawing, 324
Conductance plot, 98–99
Conduction band, 13
Conduction band edge (E_c), 14
Conductivity, 19–20
Contact enhancement, 142
Critical dimensions (CD), 177, 325
Cumulative failure distribution function, 346–347
Current sources, 230–231
CV plots:
  analysis program for, 70–71, 78–79
  deep-depletion, 57–58
  deviations from the ideal, 58–67
  equations for, 53–55
  flatband shift, 58
  high-frequency, 49–50
  low-frequency, 51–53
  moisture on wafer surface, 69–70
  normalization of, 55–57
  practical considerations in doing, 67–70
  sweep rate, 68–69

D flip flop, 200–202
Dark current, 71, 251

Debye length, extrinsic, 247
Deep depletion, 57–58
Deionized water (DI water), 105, 183
Depletion, 33
Depletion mode, 4, 99–101, 192–193
Depletion region, 21
Deposition process, 172–175
Design rules, 317–321
Diffusion:
  drive-in, 157–159
  equations of, 155–162
  junction depth, 159
  monitor of, 159–162
  predeposition, 156–157
Diffusion length, 23
Digitizing, 324
Diode equation, 24–26
Direct contact printing, 177
Direct step on wafer (DSW), 178–179, 327
Donors, 16–17
Drain, 1
Drive-in, 157–159
Dry etching, 181
Dynamic logic, 204–207
Dynamic RAM, 208–210

Electrically alterable ROM (EAROM), 215
Electrically erasable PROM (EE-PROM), 215–216
Electrically programmable ROM (EPROM), 213–215
Electromigration, 175
Electron beam (E beam):
  aluminum deposition, 174–175
  pattern generator, 179
Electrostatic protection, 104–105
Encroachment, 139
Energy-band diagram, 33–36
Enhancement mode, 4, 102, 192
Epitaxial deposition, 172–173
Epitaxial layer, use of, 274–276
ERASE cycle (*see* Programmable read-only memory)
Etch, 179–182
Etch rate, table of, 181
Etch stop, 180
Exclusive-OR function, 202
Exponential tail, 166

Failure mechanism, 345
Failure mode, 345
Failure rate, instantaneous, 351
Failure unit (FIT), 351
Falltime, 196
Faraday cup, 164
Fermi level (E_F), 15
Fermi potential, 22
Field threshold, 134
Finite impulse response (FIR) filter, 254
Fixed oxide charge (Q_f, Q_{ss}), 42–44, 60
Flatband, 33, 56
Flatband voltage (V_{FB}), 39–41, 111
Flicker noise (*see* 1/*f* noise)
Floating gate, 213, 215
Floor plan of chip, 323
Foundry, silicon, 328
Four-point probe, 160–161
Fowler-Nordheim tunneling, 215
Freak distribution, 350

Gate array, 328–330
Gate-controlled diode, 74–75
Gauss theorem, 38
Gaussian distribution of ion implantation, 166
Gaussian function, 159
(*See also* Diffusion)
Gold doping, 274
Gradual-channel approximation, 94
Graphoepitaxy, 172
Guard rings, 138, 264, 272

HCl, 153, 173
Hermetic sealing, 335
Hexamethyldisilazane (HMDS) adhesion promoter, 176
High-efficiency particulate air (HEPA) filter, 183
High-pressure oxidation, 154–155
Hillock, 175
HMOS (high-performance MOS) process, 142–143, 321
four device thresholds, 143

Impurity profile measurement, 65–67
Infant mortality, 350
Infinite impulse response (IIR) filter, 254

Insulated gate field-effect transistor (IGFET), 2
Interface trap density (D_{it}), 60
Intervening layer, implant through, 167
Intrinsic carrier concentration (n_i), 14–15
Intrinsic silicon, 13, 33
Inversion, 34
Inversion-layer charge Q_n, 39
Inverter:
  dc analysis of, 189–194
  definition of, 187
  depletion load, 102–103
  depletion-mode load, 192–193
  falltime, 196
  noise margin, 194
  resistor load, 189
  risetime, 195
  saturated enhancement load, 102, 190–192
  transfer characteristics of, 193
  transient analysis of, 195–197
  triode enhancement load, 102, 192
Ion implantation:
  advantages of, 163
  anneal, 168–172
  channeling, 166
  dose, 164–165
  exponential tail, 166
  gaussian distribution of, 166
  laser annealing, 169–172
  process of, 162–172
  range R_p, 164–165
  range tables for, 166–170
  standard deviation of range ΔR_p, 165
Ion milling, 182

Joint Electron Device Engineering Council (JEDEC), 337
Junction depth, 159

Lambda (λ):
  design rules based on, 328
  output resistance parameter, 225, 258
Laminar flow hoods, 183
Laser annealing, 169–172

Lasers:
    continuous wave (CW), 170
    pulsed, 170
    Q-switched, 170
Latches, 199-200
Latch-up, CMOS, 270–277
    prevention of, 271–275
    testing, 275
Lateral *pnp* in CMOS latch-up, 270–
    271
Layout, 317–321, 323
    automated, 327–331
Leadless chip carrier (LCC), 337
Load devices (*See* Inverter)
Load lines, 230
Localized oxidation of silicon (LO-
    COS), 139
Logic circuits, 197–207
Logic hazard, 322
Lognormal distribution, 345–347
Low-pass *LC* filter, example of, 243–
    245
Low-pressure CVD (LPCVD), 174
Low-temperature oxide (LTO), 174

Mask, 177, 326
Matching of components, data on, 239
Memory circuits, 207–216
Metal-gate CMOS, 136–138
Metal-gate NMOS, 135–136
Metal-gate PMOS, 133–135
Metal-semiconductor work function
    difference (ϕ_{MS}), 39, 42, 59
Midgap (E_i), 14
Miller effect, 138
Miller indexes, 11–13
Minority carrier, 2
Minority carrier lifetime:
    measurements of, 71–75
    reduction in, 274
Mobile ions, 62
Mobility:
    bulk, 19–20
    surface, nonconstant, 90–91
MOS (metal-oxide-semiconductor)
    transistor:
    characteristics of, 5–6
    comparison with bipolar transistors,
        7–8
    depletion-mode, 4, 99–101
    enhancement-mode, 4, 102

MOS (metal-oxide semiconductor)
    transistor (*Cont.*):
    load device, 101–103
    second-order effects, 90–95
    transfer characteristics of, 6–7
Murphy yield model, 341–342

n-channel device, 2, 51
n-well, 268–269, 275
NAND gate, 187
Neutron irradiation, 274
Nitride, silicon, 139, 210, 215
Noise margin, 194
Noise in MOS device, 226–227
Nonuniform doping, 109–111
Nonvolatile memory (*see* Read-only
    memory)
NOR gate, 187
Number of squares, 162

$1/f$ noise, 226–227
Operational amplifiers (op-amps):
    CMOS, 233–236
    NMOS, 237
Output buffers, 2, 216–218, 231–232
Oversized mask, 145
Oxidation:
    equations of, 149–155
    HCl, addition of, 153
    high-pressure, 154–155
    redistribution of impurities during,
        66–67, 153–154
Oxidation-induced stacking fault
    (OSF), 154

p-channel device, 5
p-tub, 69, 136, 267–269
Package types, 332–337
Pattern generation (PG) tape, 326
Pellicle, 327
Phosphosilicate glass (PSG), 134
Photolithography, 175–179
Photoresist, 176–179
Pinch-off point, 84
Pitch, 319
Plasma etcher:
    barrel type, 181
    parallel plate type, 181
Plasma etching, 181
Plastic dual in-line (DIP) package, 332–
    334

pn junction:
  breakdown voltage, 26–27
  built-in potential, 21
  electric-field, 26–27
  forward-bias, 22–24
  reverse-bias, 24
  zero-bias, 20–22
Poisson yield model, 341
Poisson's equation, 26
Pole-splitting capacitor, 236
Polyimide, 210
Polysilicon, 1, 138, 277
  p-type, use of, 41, 59, 136, 270
Potential well, 248
Power-on clear, 203
Predeposition, 156–157
Price yield model, 343
Programmable logic array (PLA), 212,
  330–331
Programmable read-only memory
  (PROM), 212–213
Projection printing, 178
Punch-through, 87

Race condition, 322
Radiation effects, 64
Radiation hardening, 65, 277
Random access memory (RAM), 207–
  211
Range (*see* Ion implantation)
Ratioed logic, static, 204
Ratioless logic, dynamic, 205
RCA clean solution, 182
Reactive ion etching (RIE), 181
READ cycle (*see* Programmable read-
  only memory)
Read-only memory (ROM), 211–212
Refractory material, 138
Reliability and failure models, 344–351
Resistor, MOS transistor as, 103–104
Reticle, 179, 326
Reverse osmosis (RO), 183
Risetime, 195
RS (reset-set) flip flop, 197–199
Run-out, mask, 321

Sampled data system, 241
Sapphire, 276
Saturation region, 84–86, 102

Scaling, 125–129
Schematic diagram, 322
Schmitt trigger, 202–204
Seeds yield model, 342
Self-aligned process, 138
Semiconductor physics, review of, 11–
  31
Sense amplifier, 208, 210
Shallow states, 16, 18
Sheet resistance, 160–161
Shift register, 200–202
Short-channel devices, subthreshold
  current in, 121–122
Short-channel effect, 87–88, 114–120
Shrink, 125–127
Silicides, metal, 145
Silicon:
  crystalline nature of, 11–13
  extrinsic, 16
  intrinsic, 13–16
Silicon gate NMOS, 136, 138–142
Silicon on insulator (SOI), 277
Silicon on sapphire (SOS), 276–277
Sizing, 326
Slew rate, 236
Snowplow in doping concentration, 67
Source, 1
Sputtering, 175
Square law, 85
Standard cell, 328
Standard deviation for lognormal dis-
  tribution, 347–349
Standard floor plan, 331
Static protection:
  designing in, 104–105
  Military Spec. #883B, 106
  testing, 106
Static RAM, 207–208
Substrate bias generator, 219–220
Subthreshold current, 111–114
  in short-channel devices, 121–122
Subthreshold region, 88–90
Surface potential, 245–249
Switched-capacitor filter, 239–245
  example of, 243–245

T (toggle) flip flop, 200
Tap weights, 254
Target, sputtering, 175
Threshold voltage:
  back-gate bias effect, 91–94

Threshold voltage (*Cont.*):
  calculation of, 31–39
  definition of, 34
  drain voltage, effect on, 120–121
  narrow-width effect, 122
  nonideal effects on, 39–46
  nonuniform doping, effect on, 109
  short-channel effect, 114–120
  small-geometry effect, 124–125
  ways of measuring, 95–99
Transconductance (g_m), 223–224
Transient analysis sample run, 281–283
Transmission gates, 264
Transversal filter, 254
Triode region, 81–83, 102, 192
Tristate output, 216–218

Ultraviolet (UV) light, 181, 183, 214
Undercut, 180

Unipolar characteristic of MOS transistor, 5
Unit cell, 11

Valence band, 13
Valence-band edge (E_v), 14
Vertical *npn* in CMOS latch-up, 270–271
Voltage reference, 232–233

Wet etch, 180
WRITE cycle (*see* Programmable read-only memory)

X-ray printing, 179

Yield curves, 343–346
Yield models, 340–344

About the Author

DeWitt G. Ong holds a B.S.E.E. degree from University of the Philippines, a M.S.E.E. from Stanford University, and a Ph.D.E.E. from Purdue University. He is presently program manager for Intel in telecom design engineering. He started with Bell Laboratories designing PMOS and NMOS digital as well as bipolar analog integrated circuits. Later, at Motorola, he developed a combined charge-coupled device and complementary MOS (CCD/CMOS) process and worked on advanced CMOS processes. The author of three technical papers and holder of one patent, he developed *Modern MOS Technology* from courses he taught in MOS technology and integrated circuit design both in industry and at the graduate level at Arizona State University.